PENGUIN BOOKS

# As Far As We Know

# As Far As We Know

Conversations about science, life and the universe

**Paul Callaghan and Kim Hill**

PENGUIN BOOKS

PENGUIN BOOKS
Published by the Penguin Group
Penguin Group (NZ), 67 Apollo Drive, Rosedale,
North Shore 0632, New Zealand (a division of Pearson New Zealand Ltd)
Penguin Group (USA) Inc., 375 Hudson Street,
New York, New York 10014, USA
Penguin Group (Canada), 90 Eglinton Avenue East, Suite 700, Toronto,
Ontario, M4P 2Y3, Canada (a division of Pearson Penguin Canada Inc.)
Penguin Books Ltd, 80 Strand, London, WC2R 0RL, England
Penguin Ireland, 25 St Stephen's Green,
Dublin 2, Ireland (a division of Penguin Books Ltd)
Penguin Group (Australia), 250 Camberwell Road, Camberwell,
Victoria 3124, Australia (a division of Pearson Australia Group Pty Ltd)
Penguin Books India Pvt Ltd, 11, Community Centre,
Panchsheel Park, New Delhi – 110 017, India
Penguin Books (South Africa) (Pty) Ltd, 24 Sturdee Avenue,
Rosebank, Johannesburg 2196, South Africa

Penguin Books Ltd, Registered Offices: 80 Strand, London, WC2R 0RL, England

First published by Penguin Group (NZ), 2007
1 3 5 7 9 10 8 6 4 2

Designed by Anna Egan-Reid
Typeset by Egan Reid
Printed in Australia by McPherson's Printing Group

ISBN: 978-014-300714-2

A catalogue record for this book is available
from the National Library of New Zealand.

www.penguin.co.nz

# CONTENTS

# PREFACE

Science has never been so important to us all, and yet it seems a mysterious subject to so many. Science lies behind accelerating technological change, rapid advances in medicine and changing social attitudes resulting from both. It challenges societies with difficult issues about how we use new scientific knowledge, often requiring us to face the question of what it is to be human. And yet scientists have no greater wisdom or insight regarding ethical questions. They are matters that call on the wisdom of all citizens.

And now, as we ponder the human-imposed vulnerability of planet earth, and its climate and biodiversity in particular, we see that scientific understanding is essential not only to understanding this vulnerability, but also to how we might ameliorate the consequences. With earth's human population set to rise to around 10 billion in the next few decades, and ever-increasing demands on earth's resources, our children's and grandchildren's futures may depend on ingenious solutions based on environmentally friendly technologies.

Ironically, this upsurge in the significance of science appears to have been accompanied by a retreat on the part of some to fundamentalism, at times accompanied by a rejection of science. Hostility to Darwinian evolution, and misunderstanding of its meaning, is one manifestation of this reaction. Yet evolution, as scientific insight, is not only powerful and successful, but also explains so much of the natural world that surrounds us. These new understandings about how life evolved on earth are now matched by astonishing advances in our understanding of the birth and subsequent evolution of the universe itself.

To be unaware of the ideas that lie behind evolution, cosmology and life itself is to be denied an appreciation of some of the most profoundly beautiful intellectual achievements of humanity. But it is the responsibility of science to tell its own story, and to communicate as widely and as clearly as possible why the scientific view enriches human understanding. The last decade has seen a steady growth of science communication literature through books of remarkable quality aimed at a general audience. This book is one approach to telling the story of science, through the conversational genre.

The book grew out of a series of 10- to 15-minute conversations each month on Radio New Zealand National from 2004 to 2007, in a slot intended for a non-specialist, but intelligent, audience. The subjects traversed reflected the interests of both of us, broadcaster and physicist, and were unrehearsed, nearly all live to air. In every case just the broad topic was agreed in advance. Each of us was able to bring some prior insight to what became a series of spontaneous discussions in which the broadcaster (KH) quizzed the physicist (PC), challenging the scientific perspective when needed.

Because of the colloquial nature of the conversation, the verbatim transcripts needed editing, with occasional tightening or clarification. Nonetheless, we have tried to stay as close to the original as possible and the resulting text will be quite recognisable to those who heard the original broadcasts. The shorter chapters represent a single radio conversation, but others are an amalgam of two broadcasts where the topic extended over to the second month. This book is not intended as a work of scholarship with extensive references and footnotes. However, both of us have been strongly influenced by a range of excellent books that provided background reading. We list, at the end, a bibliography that readers might find helpful, as well as references to articles or websites where readers can pursue topics further. We suggest, for internet users, that the Wikipedia website is generally accurate and informative.

There are 16 chapters in this book, on topics ranging across physics, chemistry, earth science, biology and cosmology – sufficiently broad to allow an overview of ideas involved in different branches of science.

The ordering of the topics is not the same as the broadcast order, but has been selected to provide the best flow and coherence. We chose to finish with quantum mechanics because, of all our subjects, this is the most bizarre and challenging, though we each admit to finding it particularly fascinating and enjoyable. If there is one overarching theme, it is our attempt to answer the question 'What is science?' If, on reading this book, the answer to that question is a little clearer, then we will have achieved our purpose.

We are grateful to various people who have supported this project: to Alison Brook and Andrea Coppock of Penguin Books, to Margaret Brown and Naomi Guyer at the MacDiarmid Institute, to Chris Bourke, Mark Cubey and John Barr at Radio New Zealand, to Andrea Keesey and Giselle Marion for help with transcription, to Glenda Lewis of the Royal Society of New Zealand for helpful advice and feedback, and to Dylan Horrocks for his beautiful illustrations.

Paul Callaghan and Kim Hill, Wellington,
February 2007

# FOREWORD

New Zealand's greatest scientist, Ernest Rutherford, said that scientists should be able to explain their discoveries to a barmaid. This remark, shorn of its snobbery and sexism, contains an essential truth. The essence of science can be appreciated by anyone. Science seems forbidding because of the technicalities, the mathematics and the jargon. Specialists need to master these – but it's the key ideas, not the technicalities, that really matter. You can appreciate music even if you can't play it or read musical notation. Likewise, you can appreciate science without being a scientist.

Indeed, it's a real deprivation to be unable to appreciate the marvellous vision of nature offered by Darwinism and by modern cosmology – the chain of emergent complexity leading from a 'big bang' to stars, planets, biospheres and human brains able to ponder the wonder and the mystery of it all.

Science is the nearest we have to a universal culture. The night sky has been shared by all cultures through human history – though it's been interpreted in many and various ways. It's not just the starry heavens that look the same the world over. Protons, proteins and Pythagoras's theorem are the same everywhere.

But science is more than part of our culture: it's pervasive in our lives, and technology is ever more sophisticated. Indeed the rising sophistication of everyday technology is actually a barrier. Inquisitive children in the mid 20th century could dismantle a clock, radio set, or motorbike, figure out how it worked – and even put it together again. If you owned an old car you couldn't keep it on the road without some mechanical skill.

It's quite different with the marvellous artifacts that pervade our lives today – mobile phones, iPods, and the rest. It's hard to take them to bits. If you do, you'll find few clues to their arcane miniaturised mechanisms. They're baffling 'black boxes' – pure magic to most people.

For very young children, the wonder's there – whether focused on space, dinosaurs, or tadpoles. But it sometimes fades by the time we've grown up. Youthful enthusiasms aren't always sustained.

These radio conversations cover all the key ideas of science – how science enables us to understand ourselves, our essential technology, and our threatened environment. They remind us that science is an unending quest: each advance brings into focus some new questions.

Paul Callaghan is an eminent scientist who wears his learning lightly. Kim Hill plays the part of the intelligent inquirer. Between them, they cover an immense range of topics with clarity, and even with humour. These conversations deserve to be preserved in more permanent form. This very welcome and enjoyable book will 'spread the word' to many who weren't fortunate enough to hear the original broadcasts. It deserves a very wide readership. It's a pleasure to recommend this book.

Martin Rees
President, Royal Society of London

## CHAPTER 1

# WHAT IS SCIENCE?

To myself I seem to have been only like a boy playing on the seashore
... whilst the great ocean of truth lay all undiscovered before me.
– SIR ISAAC NEWTON

*KH: So, what is science? Did somebody invent it?*

PC: Science was gradually discovered. I don't think any one person invented it. And as for what science is, there is no agreed definition. But I like one description by Lewis Wolpert, which is that it's a means of looking at the world to try to understand natural phenomena and their causes in a way that is self-consistent and corresponds with reality. The ideas of self-consistency and correspondence with reality are the two key ideas of science. The correspondence with reality means that you do experiments, you make observations and you test ideas. That might seem the most obvious thing but, actually, it had to be discovered. And the idea of self-consistency is enormously important, too, because ideas in science can't just stand alone. They have to be consistent with all we know.

*KH: When did science begin?*

PC: The origins of science are hard to find precisely, but what we do know is that following the Renaissance there was a revolution, an explosion of knowledge and of understanding what science could do. That revolution ultimately found its expression in England, culminating in the formation of the Royal Society in the 1600s. At that point you can really see all the elements of science coming together, including the role of experimental tests, the processes of publishing, of scientists gathering together to discuss problems.

*KH: This isn't some narrow Eurocentric view, is it?*

**PC:** Of course there was always knowledge, in all cultures. The Chinese knew about astronomy, the Egyptians had some knowledge about chemistry, and the early Polynesians knew a great deal about navigation. But science, in the sense of Lewis Wolpert's description, is much more recent.

**KH:** *Then what about Archimedes? I mean, there he is sitting in the bath and he realises the displacement theory. Isn't that science?*

**PC:** Exactly. Of course the Greeks had a lot of understanding and, going back to Aristotle with his ideas about nature and basic ideas of physics, they were powerful ideas in their time. But the problem was that knowledge became locked into the clever conjectures by the philosophers, and it became a canon of knowledge that one learned. It was unquestioned. Whereas in science, as we now know it, everything is questioned.

**KH:** *So does that bring us to the idea that if it's a real scientific theory it has to be able to be disproved?*

**PC:** It should be able to be disproved.

**KH:** *I don't quite understand that. Is it like black swans? In the sense that if you have a theory that all swans are white, and if you find a black swan, then that disproves the theory?*

**PC:** Yes. But this idea of 'falsifiability' is quite modern. Karl Popper expressed this. His point of view was that if a theory or idea was to be truly scientific, then it had to be able to be falsified. In other words the theory should be able to make a prediction, which we could go and test, so that if the test failed it would be possible to say that this theory was wrong. I think that's a rather narrow view of science. It wouldn't necessarily apply to a lot of science that is more descriptive or observational, like biology. It's a comfortable definition for physical sciences, but in other areas of science it's a little too narrow. I think the broader concept of self-consistency and its basis in observation is one that encompasses all of science.

*KH:* OK. So who do you think was the first scientist, as we know it?

**PC:** There were a few people, who lived at around the same time. One was William Gilbert.

*KH:* I've never heard of him!

**PC:** He's not well known. Gilbert was an Elizabethan, a physician and the first person to scientifically investigate magnetism. For example, he showed that magnetism was different from electricity, as seen in the 'amber effect' where if you rubbed a piece of amber then it would become electrostatic and attract little feathers or pieces of paper. Magnetism came from lodestone, and Gilbert showed the difference. He showed that the earth's magnetic field was able to orient a piece of lodestone so that it pointed north. He didn't know about the earth's magnetic field and its origins, but he found out the basic rules of compasses. But most importantly, he specifically said that knowledge was not to be found from the conjectures of the philosophers, but from measurement. That really was an important idea.

Of course the person we've all heard of, and who lived at roughly the same time as Gilbert, was Galileo Galilei. Galileo was the first professional scientist, a person who devoted his lifetime to using experiment, to using observation, and not just in one branch of science such as magnetism, but across a whole range – in optics, in mechanics, in astronomy. He had an enormous impact. But Galileo's problem was that he got into trouble with the Catholic Church and that dead hand of the church set back science in the early 17th century.

*KH: Is this your version of* South Park*? You're going to get us into trouble now.*

**PC:** I hope not. I think the Catholic Church takes a very different view of this now.

*KH: Oh, it's all changed now, I'm sure, Paul. But back then . . .*

**PC:** So back then, England became the obvious place for science. Henry VIII had dismantled the power of the Catholic Church in England. It was possible for science ideas to be discussed more openly. One of the other scientists to emerge, in England, was Francis Bacon, a curious man, a politician and member of parliament under Elizabeth I. He became Lord Chancellor under James I.

**KH:** *And could well have written Shakespeare's plays...?*

**PC:** Could have, perhaps... but I prefer to think not!

**KH:** *Not much scientific method on that one, though?*

**PC:** Bacon was someone who tried to systematise what science was about. He said that science was about humans interpreting and discovering nature. He had the idea that science was something that 'levelled men's wits'. In other words, if you could work out the system of scientific evaluation and observation, then it wouldn't matter how clever you were. Of course, science is not really quite like that. Creativity and intelligence are enormously important. But it was a worthwhile proposition. And Bacon had another very powerful idea which was that the 'true and lawful purpose of science', as he put it, was to enhance human life with new discoveries and new power. In other words, that science would be 'useful'. It would bring benefits to humanity. That's a very important concept that would be picked up later in the ferment of science which was to follow.

**KH:** *You mean that Galileo didn't think in terms of usefulness?*

**PC:** Of course he must have. After all, in his development of the telescope he gave one to the Doge of Venice, so Galileo was political in that he knew the technical benefits from science, as well as the value of patronage. But Bacon really saw social benefit as one of the main purposes of science.

**KH:** *So what was this ferment you referred to?*

**PC:** The scientific revolution really took off around the period of the English Civil War. The ferment of ideas around that period and the process of groups of individuals coming together and starting to talk about science, and doing science experiments together, makes a fascinating story.

*KH: And the Royal Society of London coincided with this flourishing of science?*

**PC:** Yes, it did. It was the very first scientific society in the world. It was formed out of a meeting that took place in November 1660, which was the time of the Restoration of the monarchy and the return of Charles II as king. Before that, groups of scientists, whether parliamentarian or royalist in sympathies, had put politics to one side, and had come together in meetings in Oxford and London. The move to London resulted from changing fortunes in Oxford as the parliamentarians gained control of what was previously a royalist stronghold. People were having to leave their posts and go elsewhere. The gatherings in London took place at a location known then as Gresham College, and these led to the formation of the society. The crucial factor was that this society wasn't just a group of clever scientists. It was a mixture of people with scientific interests and political connections. One in particular was Viscount Brouncker, who had a close relationship with the new king and was able to get Charles II's imprimatur in setting up the Royal Society with a charter. The Royal Society then started publication with its *Philosophical Transactions*, and thus began the new scientific literature and all the processes around that.

*KH: And Isaac Newton?*

**PC:** Newton was the person who came along later as president of the Royal Society and really professionalised it. He was the one who removed the 'interested gentlemen' from the fellowship.

*KH: The dilettantes. But surely these people were important because they brought money?*

**PC:** Oh yes. They brought patronage from the king as well, and other political connections. Very important at the beginning, but once science showed its value and usefulness, it could stand on its own. Newton was the professionaliser and was also the one who made mathematics the 'handmaiden' of science, the language of science. That enormously increased its power. And when Edmond Halley was later able to use this approach to predict eclipses, and especially the return of the comet now known as Halley's comet, in a way that was so publicly obvious, then this was the ultimate triumph in the public view. So this predictive capacity of science was established at a time when its usefulness was becoming more and more obvious.

**KH:** *Let's talk some more about that.*

CHAPTER 2

# USEFULNESS

When I find myself in the company of scientists, I feel like a shabby
curate who has strayed by mistake into a drawing room full of dukes.
— W.H. AUDEN

*KH:  How about the usefulness of science, the major discoveries or inventions of science? Perhaps you could give an entire overview of the history of civilisation in a few minutes, starting now ...*

PC:  I'll try. I thought a bit about this when I had to give a talk to 'Treasury' recently. They asked me to speak about science and its connection to wealth and economies, and so I got together a list of examples where science had made a difference to people's lives. I came up with seven or eight ideas and I'm sure anyone else could come up with a different set.

*KH:  Why don't you go through the list.*

PC:  I wanted to talk about things that came out of science, rather than direct inventions like the wheel. The big one I'd start with would be lenses and spectacles and what it meant to people's lives, that they could wear glasses. That's a big one. Another of enormous importance to New Zealand, following a little later on, was the steam engine and its descendant, refrigeration, coming out of an understanding of thermo-dynamics. It was refrigeration that totally transformed New Zealand's economy.

*KH:  Otherwise we'd still be a subsistence economy.*

PC:  The next one on my list is the invention of wireless telegraphy, and then radio, which really came out of the science of Maxwell's understanding of electromagnetism. That was really huge and obviously so for New Zealand, being connected to the rest of the world. Another example, at the end of the 19th century, was the Haber–Bosch process, the way in which you could fix nitrogen from the air to make ammonia for fertiliser. This suddenly changed the agricultural potential of the planet and made it possible to increase food supply, without the need to depend on nitrates dug out of the ground.

*KH: With good and bad effects, really.*

**PC:** Good and bad. Bad, perhaps, in that it has enabled 6.5 billion people to live on the planet. The next one on my list would be the discovery of the electron by J.J. Thomson in 1897 and then, at the start of the 20th century, Rutherford probing atomic structure, followed by all the discoveries in quantum mechanics. The modern world of electronics is completely dependent on these understandings. And then, I would say penicillin would have to be a monumental discovery. I guess the story is well known, about Alexander Fleming, but really the work of Howard Florey led to the crucial understanding.

*KH: Because Fleming didn't really know what he was doing, did he?*

**PC:** Not really.

*KH: This is a shame. Fleming's is the name we always remember.*

**PC:** I know. Often names are attached to discoveries that weren't necessarily that deserving.

*KH: You're not talking about our very own Rutherford, are you?*

**PC:** Oh no. Totally deserving there, of course! So anyway, antibiotics have got to be one of the great discoveries of the 20th century for all of humanity. Next I think I'd move on to the discovery of DNA structure, which, when combined with Darwin's understanding of natural selection, suddenly meant that we humans could see where life came from, and that, in consequence, has had an enormous impact on the way we view ourselves, and perhaps our role in the universe.

*KH: That's a big call. How do we get from DNA to our role in the universe?*

PC: Because with the understanding of DNA you get to see how natural selection works, and once you understand natural selection you understand evolution, and once you understand evolution you see that humanity is but a small part of a life process that has gone on for thousands of millions of years, and that we are not necessarily the triumphal end point at all.

*KH: No. No. We might be the tea break!*

PC: Did you see the recent newspaper report on the estimates of what would happen if humans disappeared, of how long it would take for all traces of humanity to go?

*KH: Yes, I did. Not long.*

PC: A maximum of 200,000 years. In just 50,000 years it would be hard for some alien visiting earth to discover that humanity had ever existed.

*KH: Isn't that really, really unnerving?*

PC: It is quite, isn't it?

*KH: Sic transit gloria mundi!*

PC: Well, the last discovery on my list is the contraceptive pill. It indeed came straight out of science. I've seen in my generation how the pill has totally transformed social relationships, how it has released the productive potential of half the population.

*KH: Released the productive potential by curbing the reproductive potential!*

PC: Something like that. Whether you like it or don't like it, there is a major social impact. So that's my list, but another person might pick a different set.

**KH:** *They might. Let's go back to the spectacles then, the eyeglass. The eyeglass would have been no damn good without the electric light. How on earth would we have got on if we had to do things by natural light or candlelight? Candlelight must have been debilitating and probably increased the need for spectacles.*

**PC:** Maybe.

**KH:** *Was it Galileo who developed spectacles?*

**PC:** Galileo came a bit later, in the 16th and early 17th century, and was a major part of the development of lenses.

**KH:** *So who discovered them in the first place? The Chinese?*

**PC:** Many cultures had them. In Europe they would say that they emerged from Italy, one person associated with them being Nicholas of Cusa, in 1450 or thereabouts, but even the Romans knew that certain crystals would refract the light in a way that enabled people to see better, so there was a longstanding awareness there. But a breakthrough came in the 17th century with Galileo Galilei in Italy, Hans Lippershey in the Netherlands, and with others such as Johannes Kepler. These were the people who learned how to grind the lens, to understand the shape of the lens.

**KH:** *So the first impetus was for the individual, to improve the eyesight, or was it to gaze at things in space?*

**PC:** Oh no, I'm sure that initially it was about improving eyesight. You can imagine the artisans, the skilled people, who beyond 40 to 45 years of age would lose the capacity to do skilled work requiring good eyesight, suddenly having their productive potential enormously increased. They could read longer and carry out fine craft later in their lives. It must have been a huge impetus to human productivity.

**KH:** *I wonder if, in general, our eyesight now is worse, I mean sans spectacles, than it was before spectacles were invented?*

**PC:** It has to be, doesn't it?

*KH: Why?*

**PC:** Because there would have been a natural selection against bad eyesight. Especially if you were born short-sighted, your chance of survival would have been poorer, with a greater chance of being knocked over by a horse and cart, for example, before you could breed. With the advent of spectacles, there was a smaller disadvantage. Personally, I think my eyesight got worse when I started wearing glasses, although my optometrist assures me otherwise. But then they would say that, wouldn't they?

*KH: They would. I have a sneaking suspicion that if you rely on something then your natural predisposition is to regress. But being blind as a bat myself...*

**PC:** God bless the discovery of spectacles.

*KH: Even those pink-plastic-rimmed ones that I had to wear as a child.*

**PC:** That sounds pretty dreadful. Anyway, the development of the lens led also to the telescope, and hence an expanding knowledge of the nature of planets and the universe – again, very important in learning who we are and what our place is, with obviously some impact on people's thinking. Then with the development of the microscope, with the work of people like Robert Hooke, who saw the microscopic world, we learned about micro-organisms and a whole world of tiny bacteria. All that so much extended our understanding of the natural world. The lens was huge. No question about that.

*KH: Yes. So, steam engine and refrigeration. Is this because it's particularly refrigeration for us, in New Zealand?*

**PC:** Once the steam engine was there, the ideas of thermodynamics drove the discovery of refrigeration. Most people think of James Watt in association with the steam engine, though a Frenchman, Denis Papin, working with the German Gottfried Leibniz, showed quite a bit earlier how the idea could work. But Watt developed and patented the steam engine. The thermodynamic idea was that you could run a heat engine backwards, to make a refrigerator; instead of using heat to make work, to use work to extract heat. A Scotsman, William Cullen, demonstrated refrigeration principles in the 18th century, but the refrigerator, as we know it, wasn't developed as a practical tool until the middle of the 19th century, by Jacob Perkins and Carl von Linden. In the 1890s refrigeration was introduced into shipping and then the New Zealand economy gained a boost.

**KH:** *On the back of refrigeration.*

**PC:** Yes, suddenly creating significant wealth for some and leading to some of the grand country homes of that period, for example in the Wairarapa and Rangitikei districts of the North Island.

**KH:** *Those were the days. The telegraph?*

**PC:** Electronic signals down wires, tapping Morse code – that was around from the early 19th century. But I'm more interested in wireless, the development of radio, using electromagnetic waves to transmit information. Again, from New Zealand's standpoint, it made a huge difference to our connection to the world.

**KH:** *Better than shoving notes down the bellies of frozen animals!*

**PC:** I guess so. An interesting form of mail.

**KH:** *We used to write letters in the old days!*

**PC:** We did. We even had radio entertainment when I was a child.

**KH:** *Yeah, but you probably had to pedal it!*

**PC:** But anyway, the thing is that radio communication really came out of science, out of James Maxwell's understanding of electricity and magnetism, expressed through his famous four equations, which then predicted electromagnetic waves, explaining light, but also the wider electromagnetic spectrum, with radio waves part of that. Then Heinrich Hertz actually showed how to generate radio waves, and Nikolas Tesla, the Hungarian-American engineer, was the first to demonstrate radio transmission, and Marconi showed how to use it for communication and commercialised it. Interestingly, our own Ernest Rutherford was heavily involved with radio waves in his early research work as a Master's student at Canterbury University in the 1890s. He developed a detector of radio waves around 100 times more sensitive than any other at that time. He was distracted when he went to Cambridge and ended up working on alpha particles, with some success!

*KH: And so atomic physics and electronics and so on?*

**PC:** Yes, no need to point out the value of that. So let's jump to Fritz Haber who found out how to use a catalyst to capture nitrogen from the air to make ammonia. With ammonia, one could produce fertilisers artificially.

*KH: And before that we relied on legumes?*

**PC:** Or nitrates dug out of the ground, for example in Chile. The other intriguing thing about Haber is that he also learned how to make high explosives and so became a controversial figure, often blamed for the terrible use of high explosives in the First World War. Ironically, as a Jew, he was hounded out of Germany in the 1930s, having previously made such an enormous contribution to German agricultural, industrial and military capability. A tragic character who made a huge impact on the ability of the world to feed itself.

*KH: Of course, that's not so highly regarded now. We're a bit smug about our ability to feed ourselves, and we see that all that nitrogen is causing huge environmental problems.*

**PC:** Exactly. Let's finish with the contraceptive pill. Gregory Pincus and Carl Djerassi discovered how to do this, how to synthesise oestrogen and progesterone, and that became available as a product in the 1960s, with largely excessive doses I might say, causing all sorts of side-effects. Now, of course, much smaller doses are used.

*KH: Carl Djerassi is a playwright.*

**PC:** Yes, a playwright. He and Roald Hoffmann, another chemist, are famous for writing the play *Oxygen*.

*KH: I wonder how much the '60s was a consequence of the contraceptive pill being available at that point. Which is cause and which is effect?*

**PC:** You mean, would the pill have been developed had there not been a prior change in social attitudes?

*KH: Yes. By the way, coming back to DNA, do you think that understanding the way it replicates, as a double helix, was the important thing?*

**PC:** Partly, but also our understanding of the genetic code. Let me give one telling example and that is our understanding of the insignificance of differences in genetic code between different races, how small they are compared with normal human genetic diversity. We now see the whole of humanity as an interconnected family. That's surely incredibly important.

*KH: It is. Well that's one list. And I'm sure others will have different ideas, but that covers the field pretty well.*

## CHAPTER 3

# THE GLORIES OF COLOUR

Any colour, so long as it's black.
– HENRY FORD

**KH:** *Light and colour – hello, what is light?*

**PC:** Light is a part of the electromagnetic spectrum. It's a wave and it's caused by little electric fields oscillating backwards and forwards in space.

**KH:** *So when it's dark there are none of those?*

**PC:** I guess that's right. When it's dark there is no light.

**KH:** *No colour and no light?*

**PC:** Yes.

**KH:** *Because light is the same as colour?*

**PC:** Light has colour as part of it, and light is just that part of the spectrum that our eyes have evolved to be sensitive to. It's a very natural thing that our eyes have this sensitivity because it's the part of the electromagnetic spectrum where the sun is most intense. Naturally life on earth would have evolved with sensitivity to radiation from the sun at the most intense part of the sun's spectrum.

**KH:** *So what does spectrum mean?*

**PC:** When I talk about spectrum, I mean the length of the waves. The spectrum is the range of wavelengths. We can talk about waves of different lengths between crests, just as an ocean wave has a certain distance between crests.

*KH: Long wavelength, long distance?*

**PC:** Exactly. So, for example, radio is carried by electromagnetic waves. People listening to this programme on AM radio will be receiving radio waves with a wavelength of about 500 metres, but if they are listening to it on FM radio, then they are receiving a three-metre wavelength. But light is a very short wavelength, in the range of 400 billionths of a metre to about 700 billionths of a metre. We more conveniently call that 400 nanometres to 700 nanometres. That's the range.

*KH: I can't make the connection between light and colour.*

**PC:** Well, the different wavelengths determine the colour. So if it's 400 nanometres it's very blue and if it's 700 nanometres it's right up at the red end. Green would be at about 550 nanometres.

*KH: And on either side there is ultraviolet and infrared.*

**PC:** Yes, but we only see the visible part, the part of the spectrum we call light. Our eyes are able to detect the light and distinguish the colour using the rods and cones.

*KH: Oh, those rods and cones!*

**PC:** Rods and cones. There are three different types of cones, sensitive to different parts of the spectrum, so there are long-wavelength cones, mid-wavelength cones and short-wavelength cones. And those different types give us different sensitivity to different wavelengths.

*KH: And if you are colour-blind, you are deficient in one of those?*

**PC:** Yes. Most people are trichromatic, but eight per cent of males are dichromatic or even monochromatic, which means they have only two types or only one type of cone. But a very few females are tetrachromatic – they have four types of cones. It's very rare and it's only females.

*KH: So what happens? What do they see?*

PC: They have an enhanced sensitivity to colour. They distinguish colour shades that no one else would see.

*KH: Ah . . . but how would they know?*

PC: How would they or we know? Colour, after all, is a perception thing.

*KH: We used to sit in the back of the bus, when we were at school, and say, 'How do you know that you're seeing what I see when you see red?'*

PC: Exactly. It's also a cultural thing. For example in Mandarin – at least in older Mandarin – I think there's a word, *qing-se*, for green and for blue, in the sense of different shades of the same colour. In Russian there are two words for blue – one for dark blue and one for light blue. So it is cultural. And it's a psychological phenomenon.

*KH: Are you sure that the Chinese won't object?*

PC: I don't want to offend a billion Chinese. Perhaps it is safer to say that the Welsh have one word, *glas*, for green and blue. There are only about a million Welsh, I think.

*KH: Now you've really done it! So you could have a world without colour, with exactly the same sort of light, but it's just that our eyes are inventing it?*

PC: Well, our cones can discriminate between different wavelengths, but how we perceive colour, that's a matter of how the brain interprets. We all have a world without colour when the light gets very dim. You will notice at night-time we hardly see colour at all and that's because the cones, which are sensitive to colour, aren't sensitive to very low levels of light. That's when the much more sensitive rods kick in. The rods are about ten thousand times more sensitive than the cones and at night-time, when the light levels get very low, we don't see colour any more. We just see black and white and grey, because now we depend on the rods to see, and so it becomes like black and white photography. We are all colour-blind when it's very dark.

*KH: Is that why stars have no colour?*

**PC:** No, stars are actually quite bright and we can distinguish their colour. For example, a star like Betelgeuse, in the constellation of Orion, looks distinctly red. The colour of stars depends on their temperature. Whitish stars are about the same temperature as our sun.

Actually the light levels we can detect are very small, because the sensitivity of the rods is quite remarkable. There is a lovely measure of that. The least bright object the human eye can see is one candle flickering 20 kilometres away. Isn't that amazing?

*KH: Well you might be able to . . .*

**PC:** No really, the average person can do that, if they allow their eyes to become accustomed to the dark, which takes about 30 minutes. If there is a low level of light, no 'light pollution', then there are these little chemical changes in the rods that make them more and more sensitive after about 30 minutes. We can pick up something like 500 photons – little lumps of light – per second, coming into our eyes.

*KH: So, coming back to colours, are they real or something we invent?*

**PC:** Something we invent, in the sense that the way in which those cones operate will be different from person to person. So the perception may be different from person to person.

*KH: But is colour objective or subjective, in a scientific sense?*

**PC:** Partly subjective, in our perception, the way that taste is subjective.

*KH:* *Well, no, that's not true because you can find chemicals in food that make identifiable impacts on your taste buds, so you can say, 'This is sweet. This is sour. This is salty.'*

PC: Yes, there is also an objective aspect to colour in the sense that different wavelengths can be assigned to different colours, but that doesn't give us the full range of colour. By adding together different amounts of primary colours, red, green and blue, you can make the entire gamut of colour that the eye and brain can perceive, which is about 10 million.

*KH:* *Ten million?*

PC: We can differentiate 10 million different colours, all the different hues. But what that means in terms of perception, what that colour feels like, what does blueness mean, that's a matter for psychology. But we can define colours, in terms of mixtures of the primaries. For example, a typical modern computer screen can generate about 17 million different colours, more than the human eye can see. It's done by mixing different amounts of red, green and blue, or by using a different scale of hue, saturation and intensity. Whichever way you do it, you need three elements, because we have three types of cones. The artist will create colour for the canvas by mixing their primary colours. But when you are mixing colour with paint, it's slightly different, because the pigments in the paint remove colour from the white light which illuminates the surface, an effect known as colour subtraction. The light that comes back is missing some colour and that's what is determining the hue. Hence the mixing process uses a different set, cyan, magenta and yellow. The best results for a complete range are obtained by including black as well, as a fourth element. Your ink-jet colour printer uses these four in colour subtraction.

*KH:* *This woman that you mentioned, the one who is tetrachromatic. When she looks at the rainbow, does she see more than seven colours, do you think?*

PC: She would see more colours, but of course the rainbow is not just seven colours. It's a continuum of colour.

*KH:* *Yes, it is, it's seven; it's red, orange, yellow, green, blue, indigo, violet.*

**PC:** 'Roy G Biv' I was taught! Isaac Newton chose seven colours because seven has special magical properties, like seven days in the week, seven deadly sins, or seven notes in the musical scale. It's really a continuum but you can identify seven colours if you want.

*KH:* *But you could have 14?*

**PC:** You could, and if you were Russian you could have different words for the different blues! And it's worth remembering that the spectrum we see in the rainbow doesn't make all the possible colours. The rainbow has the pure spectral colours, but if we add different combinations of those colours we can make non-spectral colours, like shocking pink or brown, and in nature around us, in plants and butterfly wings and in the landscape, we do see these non-spectral colours.

*KH:* *What did Newton do with his prism?*

**PC:** He passed white light through a prism, which is a triangular-shaped piece of glass, so that when the light came in it wasn't entering 'normally' or at right-angles to the surface, but obliquely. As it comes in obliquely, the light bends, it refracts at the surface, it changes angle, and the amount by which the angle of the light path changes depends on the wavelength, that is, the colour. So the light gets split up into its separate spectral colours – the rainbow in other words. And that's how Newton realised that white light was actually made up of all those colours of the rainbow joined together. Cleverly, he proved that by recombining all the rainbow of colours using a second prism, and it came out white.

*KH:* *Is that what he intended to do?*

**PC:** Of course, we don't know what he intended to do, but 'after the fact' this is how he explained it. This was a real insight. Of course, people knew about the rainbow and the separation of colours long before Newton, but it was Newton who showed that it was a refraction effect.

**KH:** *What's fluorescent light, then?*

**PC:** Fluorescence is where you illuminate a material with light, or some other radiation of some wavelength, and it subsequently glows with a different colour, at a longer wavelength. You can sometimes see this with flowers, particularly in New Zealand where we have very high ultraviolet levels, and sometimes the flowers look intensely bright. What's happening is that ultraviolet, which has a shorter wavelength than visible light, is illuminating the surface of the flower and exciting electrons in the atoms in the flower's pigments, which then de-excite by dropping to lower energy levels. These energies released are smaller than the excitation energy, and so correspond to longer wavelengths than ultraviolet – that is, in the visible part of the spectrum. The colour always shifts down the spectrum in fluorescence. Those ultraviolet lamps in bars and nightclubs that appear black make our clothes glow brightly in the visible part of the spectrum. That's fluorescence.

**KH:** *As you say, we have a lot of ultraviolet in New Zealand.*

**PC:** Yes.

**KH:** *But it doesn't just make the blue of flowers brighter, it makes all flowers brighter?*

**PC:** I think so. Fluorescence doesn't just have to involve ultraviolet. But whatever colour is used to illuminate, the fluorescence will be at a longer wavelength, so the colour is shifted. Fluorescent light bulbs also work via ultraviolet. A current is passed through a gas containing mercury vapour, the atoms are excited and generate ultraviolet light, then that light falls on the inside surface of the glass which is coated with a phosphor material which in turn fluoresces by absorbing the ultraviolet and then releasing light of longer wavelength at a particular colour in the visible part of the spectrum.

**KH:** *Coming back to the rainbow. What is that?*

**PC:** Actually, the rainbow is one of many wonderfully colourful atmospheric phenomena that are so much part of human culture, history and even religion.

**KH:** *So what are the legends of the rainbow?*

**PC:** For the Chinese, the goddess Nüwa used five different-coloured stones to fill cracks in the sky. In the Maori legend, the rainbow separates Ranginui and Papatuanuku. Every culture has its stories.

**KH:** *And what causes the rainbow?*

**PC:** It's caused by sunlight being reflected back from droplets of water. Those droplets could be due to rain falling in the distance, or from the fine spray of a garden hose. The light illuminating the droplet goes inside and is reflected off the back surface.

About 300 years ago, René Descartes, a French mathematician, worked out the geometry of this by drawing little rays going into and around the interior of the raindrop. He realised that if you followed the paths of the different rays, they would tend to come out bunched up at a particular angle to the incident light, and that angle is 42 degrees. So what it means is that if you have the sun behind you and there are raindrops falling in front of you so that the sunlight reflects back, then at what's called the 'anti-solar point', the point where your head's shadow would be in front of you and directly opposite the sun, if you drew an arc centred on that point, of 42 degrees radius, then that would be the primary rainbow. The primary rainbow is the brightest one we see. It needs the sun to be less than 42 degrees above the horizon. If it is more than 42 degrees above the horizon you will never see a rainbow, because that rainbow would be entirely below the horizon. If the sun were exactly on the horizon below you then that would give you the biggest rainbow, a perfect semicircle. So, the amount of rainbow we see depends on the altitude of the sun.

**KH:** *So the sun's rays have to make a 42-degree or less angle?*

**PC:** It's the raindrops that bunch up the rays and bring them back at 42 degrees to the incoming direction. That's why you see the bow. It doesn't matter if they are big droplets or small droplets, it's the same – a simple geometric effect of spherical droplets. You have to draw the rays to see this. There is another rainbow at 51 degrees, and that's the secondary rainbow. We often see a double rainbow, the secondary one being a bit weaker. In fact there are even more, higher-order rainbows, all in different parts of the sky. It turns out that the third-order and fourth-order bows are behind us, around the sun, and very hard to see.

*KH: Because the sun's so bright.*

**PC:** Exactly. But sometimes we can see the fifth- and sixth-order bows in front.

The other thing we notice is that the ordering of the colours is different in the bows. In the primary bow the blue is on the inside and the red on the outside, whereas in the secondary bow it's the other way around.

*KH: Oh dear, I expect that there is a very complicated explanation for that?*

**PC:** Not really. Isaac Newton realised what it was that made the colour, and Descartes worked out what made the bow. The colours are separated because of refraction, as the light goes from the air into the water; and, coming out, from the water into the air, the light bends by slightly different amounts depending on the colour. And when you look at the paths that the light can take through the spherical droplets of water, it just so happens that on the primary bow the blue comes out inside, with red on the outside, but in the secondary bow, it's the other way round. It's just the way the geometry shakes out.

*KH: So what about these other atmospheric effects you mentioned?*

**PC:** There's a wonderful book about all that by Robert Greenler called *Rainbows, Halos and Glories.* The book also talks about something called the green flash.

41

*KH: They sound fascinating. One of our listeners sent us a picture of a glory. What's the difference between a halo and a glory?*

PC: Many different names here: a gloriole, an aura, a nimbus, and a heiligenschein. The halo, to use the scientific definition, is a ring that appears around the sun when there are ice crystals in the air. In New Zealand you would only see it through very high cirrus cloud where there were ice crystals. It's much more common in colder climates and especially so in Antarctica, where it is possible to see the most spectacular haloes. They are caused by light being reflected not by raindrops, as in the rainbow, but by ice crystals. Ice crystals are complex, generally with a hexagonal shape. They have facets. And those facets produce distinctive patterns. So in Antarctica, you can see this big ring at 22 degrees around the sun.

*KH: The ring around the sun.*

PC: The ring around the sun – 22 degrees is about the angle subtended between your forefinger and your thumb, if you stretched your arm and hand out. But the crystals can make other patterns as well. And these patterns depend on whether they are pencil-like or flat crystals. Flat crystals, falling gently in the air, can produce sundogs, or parhelia – little false suns out to either side.

*KH: Ah, sundogs.*

PC: And they can also produce something called sun pillars, a column of light rising up, caused by the ice crystals reflecting light. Amazing phenomena, not seen in New Zealand's latitude, but which can be seen in the Arctic or Antarctic.

*KH: I don't know the etymology of the word, but presumably the halo was the thing that went around angels' heads.*

PC: There are many legends. For example, in the fourth century the emperor Constantine apparently saw what he thought was a cross, but was most likely a sun pillar with sundogs either side, and that's apparently what converted him to Christianity. There are many religious connotations around these effects. Where the halo is associated with the human head, the scientific term for this is a glory. A glory is something slightly different. Most of us would have seen a glory from an aeroplane. If we had been flying in an aircraft, and looked down and seen the shadow of the aircraft on a cloud layer below, we may have seen a ring of coloured light around the shadow.

*KH:* *The picture sent to us shows the shadow of a Boeing 737 crossing the Tasman, in mid-crossing, surrounded by a glory, and the shadow of the aircraft is on top of dense cloud layer. Is it because I am looking at it that there is a shadow?*

*PC:* It's because your head shadow is where the plane shadow is. Of course, the religious connotation for this came when people travelled up to high mountains. If the person makes a shadow on clouds below, then you see a glory around that shadow of your head. It was first recorded by the Chinese in the yellow mountains, Huangshang.

*KH:* *Not from a Boeing 737 I imagine!*

*PC:* No. And it was also seen in Germany, up in the Brocken Mountains, and so called the Brocken spectre. It was seen from high up, above the clouds looking down on the cloud tops around the shadow of the person viewing. It has the religious term halo, though the scientific term is glory. A glory is caused by light scattered back from very tiny droplets, and unlike the rainbow, where the light that returns is reflected by the drops, for the glory, the scattered light is diffracted. Diffraction is something that happens because of the wave nature of light. The glory is not a big 42-degree arc, but a ring much closer in, at about five to 10 degrees, depending on the size of the droplets. So the glory is a ring of light quite close in around the shadow of the head.

*KH:* *Glory is a very nice word, isn't it?*

*PC:* It's a beautiful word.

*KH:* *Very unscientific.*

**PC:** Ah, science has lots of beautiful words.

*KH:* *So what's the green flash?*

**PC:** The green flash is something that few people have seen. It's a remarkable phenomenon that occurs just at sunset. Just for a short time, less than a second, you see a flash of green light as the last part of the sun's orb drops below the horizon, and especially over the sea. And you need special conditions for that. The reason that it happens is because the air refracts the light, and the spectrum of the sunlight – the blue, the green, the red – have their paths refracted, or bent, by different amounts. The bluish part is the last to set, but so much blue is scattered out of the path of light from the sun to our eyes by the air – (that's why the sky is blue) – that the green turns out to be the last colour in the setting sun that is predominant.

The other thing you need for a really good green flash is a miraging. You need to have a temperature inversion over the sea surface – a layer of cold air under a layer of warmer air. That produces a mirage that stretches out and delays the orb dropping below the horizon, leaving the greenish part suspended there for a moment.

*KH:* *Have you seen it?*

**PC:** Yes, very clearly from the island of Oahu in Hawaii.

*KH:* *By accident or were you looking for it?*

**PC:** No, I deliberately looked. I waited there and watched the full sunset over the sea. Actually I always look for the green flash when I am outside in a sunset over a western sea. I've looked hundreds of times. But the Hawaii one was the best, by far.

*KH:* *Life is too short, Paul.*

**PC:** I'll keep looking though!

*KH: Has the green flash got anything to do with Aurora Borealis or Aurora Australis?*

**PC:** No, aurora are completely different phenomena. They are caused by the solar wind, a stream of ionised particles coming from the sun, hitting the earth's magnetic field and being directed in a stream towards the poles. As they come down through the atmosphere they excite atoms of the upper atmosphere that then de-excite by emitting light. That's what causes this amazing flickering of light overhead, in regions nearer the poles, as the ionised stream is drawn down to the magnetic poles. You can sometimes see the Aurora Australis looking south from New Zealand.

*KH: I saw the Aurora Borealis in Norway once.*

**PC:** I did, too. In Trondheim.

*KH: That was pretty amazing. It makes you think that there must be something 'going on'.*

**PC:** You can understand how all these religious ideas came about because of these powerfully beautiful phenomena. But they all have a simple physical explanation.

# CHAPTER 4

# A LITTLE OF TECHNOLOGY
# AND THE DIGITAL WORLD

Technology ... the knack of so arranging the world
that we don't have to experience it.
— MAX FRISCH

*KH: So how do we get the colour in a TV or a computer screen?*

PC: In the old-fashioned cathode-ray tube TV or monitor, as opposed to the modern flat-screen liquid-crystal or plasma type, there are three electron guns in the back of the tube, each one driven by a current representing the intensity of red or blue or green. There's no colour in the electrons, of course, but they hit the inside surface of the screen where there is a matrix of little dots of phosphor material, and when electrons hit one phosphor dot, they will cause red or green or blue light to come out, depending on which type of phosphor material is in the dot, by much the same process of fluorescence I have spoken of earlier.

*KH: Because of the nature of the phosphor?*

PC: Yes, very special materials like yttrium europium oxide that makes red, or zinc sulphide doped with silver or some other metal to make the green or the blue. And the marvellous thing about the television set is that the three electron beams have to be perfectly aligned to hit their correct phosphor dots. Sometimes a special shadow mask is used to assist that. It's very precisely engineered.

*KH: Is it dramatically different with the flat-screen TVs – the plasmas or whatever?*

PC: There is no electron gun with plasma or liquid-crystal display, or LCD, screens. What you have is a flat screen made up of many picture

49

elements called pixels and each one is individually driven by a current, aimed specially at that pixel alone. In the plasma screens each pixel is like a little fluorescent tube, a little fluorescent light bulb, but with the right phosphor to make the right colour, so you need at least three little light bulbs for each pixel so that you can generate any colour in it.

*KH: How many little glass tubes?*

**PC:** Millions of them, and all addressed independently by electrical currents travelling vertically or horizontally across the flat matrix that makes the back of the screen. Where signals intersect, a pixel is illuminated. This modern technology allows larger, flatter screens. The problem with the old cathode-ray tubes, apart from being heavy and bulky, was that you couldn't make them very wide because it is hard to bend the electron beam too far to the left or right.

*KH: So what's the difference between liquid-crystal and plasma screens?*

**PC:** LCD is the same concept – a flat screen – with individually addressed pixels but with light and colour generated in a completely different way. It's a bit more complex. It uses the fact that liquid crystals can twist the polarisation of light and hence determine how much light gets through a Polaroid film in the front. Separate liquid-crystal elements are needed for each colour in each pixel, with filters on the front imparting the colour to the light. LCD screens are less heavy than plasma and have intrinsically better definition, but they are not perhaps quite so bright as plasma at present.

*KH: So which is best?*

**PC:** It's a battle between the technologies. My guess is that LCD will win, because that 'mechanical lightness' is important. But then newer technologies based on conducting light-emitting polymers will probably take over.

*KH: And what do you have?*

**PC:** I've just got an old-fashioned cathode-ray tube.

**KH:** *That's quaint. I thought that you might say, 'I just have radio'.*

**PC:** No, really, I'd love to have one of these fancy flat screens.

**KH:** *Well, that's how it makes the colour on the screen, but how is colour transmitted to drive the TV set?*

**PC:** It's all about to change now, but let me describe the analogue TV – the type that most of us have used for 40 or 50 years, but which is being replaced by digital. The way it works is also kind of historical. If you go back in the past, to the introduction of colour television, in those days everyone had black and white TVs. It would have been an unpopular thing to do to introduce colour TV if the signal couldn't also be interpreted by all the black and white sets that people owned. So colour had to be brought in in a way that still made it possible to see the picture on a black and white set. What was done was to add another signal to the television transmission. There was already the intensity, a grey scale running from black to white, being transmitted in the electromagnetic wave received by the TV set. That was called the luminance signal. To that signal was added the chrominance signal. The chrominance contained both hue and saturation information. Remember, you always need three numbers to determine a colour and its brightness. Red, green and blue; or, in this case, luminance, saturation and hue.

**KH:** *Numbers, you say?*

**PC:** Numbers that say how much of each you need. In the transmission those numbers relate to the properties of the electromagnetic wave carrying the signal and, in the TV set, the strength of the electron currents. So luminance, saturation and hue are the three quantities on our palette; there always have to be three. Remember the three cones in our eyes? Luminance says how bright, hue says whether it's reddish or greenish or whatever, and saturation says how washed out or intense the colour is.

Luminance was perfectly good for those who just had the black and white set. So added to the old luminance signal was the chrominance, and this electromagnetic wave component had hue and saturation separately embedded. In the case of the saturation, the information was embedded through the amplitude of the wave, how big the wave was, while the hue information was contained in the phase of the wave, how much the crests in the wave were advanced or retarded in time.

*KH: So what about digital TV? Digital – what does that mean anyway?*

**PC:** I'll come back to digital TV. Let's just focus on digital in general for a bit.

*KH: OK. Give us the high tech!*

**PC:** Well, the world that we live in now is very much a digital world. Actually, the amazing thing about today is that if you talk to the average 18-year-old you'll discover pretty quickly that in their own lifetimes they've seen so many technologies just come and go. They've seen film taken over by the digital camera. They've seen tape music being replaced by CDs and they have seen the video being replaced by DVD. For young people today it's quite normal to see this rapid change in technology, and it's accelerating all the time in this digital age.

*KH: What does digital actually mean?*

**PC:** It's to do with the way we record things. The old way was called analogue because it had a direct representation of the way nature is – an analogy if you like. So, for example, in the TV case, the bigger the amplitude of the chrominance signal, the greater the colour saturation. Or in audio recording, the bigger the oscillation amplitude of the magnetism on the cassette tape, the louder the sound, and the faster the oscillation, the higher the frequency, or musical note. By contrast, in the digital world we actually write down numbers.

*KH: Analogue. Is that the same root? Analogue and analogy?*

**PC:** Yes, it is. So if you think, for example, about what music or what a moving picture is ultimately comprised of, you're going to be talking about some property that varies with time. In the case of the sound that we hear with our ears it is the air pressure at the eardrums that varies with time, and in the analogue case it's been stored on a magnetic tape in the form of magnetisation that varies with time and that is run through electronically to obtain an electrical current that drives the speakers. The problem with that approach is that if you store your information that way, then every time you copy, it can get a little degraded because you can't get a perfect copying every time you reproduce it. And as you run the storage medium again and again, it degrades a little each time. Also, little fluctuations of the quantity stored or transmitted, in magnetism on a tape, or electric fields in a radio wave, add a background hiss called 'noise'.

Digital avoids all that. By digital you turn that variation with time into a set of numbers that correspond to a value of the quantity, whether it be colour hue or audio amplitude, at each point in time of a clearly defined and pre-assigned set of equally spaced intervals. This way you can read that data using computer technology and use the numbers to drive the necessary electrical currents.

And here's the point. The numbers can be written down exactly. Even if the storage or transmission medium is a little degraded, then so long as we can still read the number, we have lost no information. Fidelity is retained. And we can make the information contained as perfect as is allowed by the precision of the numbers we choose to use and the fineness of the time intervals we choose to break up the oscillation. That degree of perfection can be summarised by the word 'bandwidth'. But the fidelity of transmission is always 100 per cent.

So, that's essentially what the digital technology does, except that the numbers used are binary numbers – in other words, written in terms of sets of ones and zeros, because that's the way computers and electronic devices can store information. Digital means that we have the opportunity to reproduce with perfection. When music is recorded in the studio, the analogue signal, the electrical current from the microphone, is immediately turned into numbers using a device known as an analogue-to-digital converter. Every time that recording is reproduced or transmitted it can be done so with perfect fidelity. But there are other advantages as well: you can actually encrypt information, you can compress information, so you can store data a lot more effectively and efficiently if it's digitised.

Digital TV is the next step, and this change is happening around the world right now.

*KH:* *Is it digital that enables people to watch something whenever they want to and stop it whenever they want to?*

**PC:** The use of the internet? Well, there are other technological reasons too. Part of the problem is that as we improve the quality of the way we store movies on digital media, like the DVD, we do reach a limitation of memory size. Eventually having a hard storage mechanism like that isn't the way to go, as we the customers want more bandwidth and more perfect and precise pictures. The best way to store is on computer memory and, best of all, someone else's computer accessible by the internet! And as the internet generation increasingly want whatever they want right here and now, the obvious way to go is to download from the internet. The digital age makes all that possible.

*KH:* *Yeah, I know, I know . . . you say improve, but it seems to me that in previous times improvement came more from consumer demand . . .*

**PC:** Yes.

*KH:* *But now people are going 'We are all right now', but . . . no, no you need a new thing . . . you need a phone that clips your toenails.*

**PC:** I know, I know. A lot of the technology we have today seems unnecessary but I can remember the argument about CDs and you'll remember, Kim, the aficionados of vinyl . . .

*KH:* *I'm sure I won't, Paul!*

**PC:** Those 'audiophiles' who said, 'Oh no, you've got to have it on vinyl – it doesn't sound the same on CD.' You never hear that any more.

Interestingly, there was one particular CD in 1985 that transformed the whole business of putting music on CD and that was *Brothers in Arms* by Dire Straits. Until then the CD had been a rather peculiar minority interest. *Brothers in Arms* was such a success because it was possible to have longer music tracks instead of the regular three-minute periods. You were able to store more music on a CD and that made that particular production a huge success which really helped the CD industry to take off.

The CD music quality is really superb – the quality is certainly superior to anything tape can do or anything that vinyl can do – so this is a technology that really improved the way in which we can enjoy music, just as the DVD technology has improved the quality of the way we can watch movies. DVD is better quality than a video because you are dealing with a perfection in copying, at least at the level of the digital resolution you decided to go for on that movie, and because you have other information on there as well, like subtitle tracks.

*KH: But soon DVDs will be looked backed upon as very primitive when we have ... what?*

PC: Well, the next stage is high-definition television, which is going to have greater numbers of pixels, larger screens for the people who happen to like to have home-entertainment systems, for example. The amount of information that starts to get stored in that type of technology will exceed the limits of present DVD technology, especially as people demand better and better resolution, sharper pictures with a greater range and trueness of colour.

*KH: Limitations why?*

PC: It's to do with how much information you can store on a disc. You know the digital versatile disc or DVD stores about seven times as much information as a compact disc or CD. CD is no good for movies. We use it for storing our jpeg files from our digital camera.

*KH: So what's the difference between the CD and DVD?*

**PC:** A DVD is a CD that's got a higher density of storage and uses a different colour laser to operate with shorter wavelengths, so that you can work with finer, more detailed structure on the disc than you do with a CD. I mean, the CD is in itself an extraordinary technology because there is something like five kilometres of tracks and those track widths are just a couple of microns, or thousandths of a millimetre. We store about 700 megabytes on a CD. But the DVD stores much more, especially if it's a multi-layered DVD. But even so, if you want to have a movie of very high quality, you start to run up against the limits of that technology too.

*KH: So what's next in terms of user pleasure?*

**PC:** Well, if I knew the answer to that I'd be a very wealthy man, which sadly I'm not. But we are living in a more colourful world. I think we're going to see bright visual displays around us. We will see more animated advertising, great liquid-crystal displays replacing billboards. And with new conducting polymers making all that cheaper, it's going to be a colourful world indeed. I imagine even clothing is going to be rippling with images in real time. Kind of unnecessary, I'm sure.

*KH: Kind of unnecessary?*

**PC:** I'm sure if it's produced someone will want to wear it.

*KH: Yeah, but what's necessary? We have lost the idea of what's necessary, haven't we?*

**PC:** We have, but if you think about something like the digital camera, I think that's improved our lives in many ways.

*KH: 'Why?', she asked bitterly!*

**PC:** Well, suppose you've got children growing up. You do want a record of them in the early stages.

*KH: Hello? What was wrong with the old camera?*

**PC:** The old camera was a pain. I mean ...

*KH: Admittedly, you had to have had those really, really bad photos developed before you knew how bad they were.*

**PC:** And if you go back and look at the pictures taken 10 or 15 years ago of your children, they've started to fade, the colours have started to go off. And if you can find the negatives to produce prints again, even they have started to deteriorate. And so when you look at generation upon generation these records gradually deteriorate. What we have now in the digital world is a medium where there's no deterioration, where we can keep in perfection the images we take of our children or parents or whatever.

*KH: On the CD?*

**PC:** On the CD, in the computer, whatever. That's it exactly. I think that's an improvement. I think the ease with which we can take a photograph is great. It's so nice to be able to take a photograph, look at it and say, 'No, that didn't come out so well, I'll take another one.' Remember the trepidation when you went to the pharmacist to pick the prints up and you were so disappointed when you saw some of the pictures hadn't turned out well?

*KH: I know, I know ...*

**PC:** Surely that's an improvement in our lives.

*KH: I know. Strangely, however, I just think it's part of the warp and weft of life to not be able to control what photos are like.*

**PC:** I think there is plenty of opportunity for failure in life in other ways, no matter how perfect our digital cameras may be. We can screw up in all sorts of ways, in every sort of thing we do. So I don't think there is any lack of opportunity for life to have the warp and weft of complexity and failure and difficulty.

*KH: Are you sure? I mean, I know it sounds self-serving, but I'm in favour of serendipity and I just wonder whether we are erasing any possibility of it with all these foolproof methods of this and that.*

**PC:** I think we create a new serendipity. I have a friend who is an artist who uses digital photography and special software to produce extraordinary images, with superposition of originals to make a new way of visualising the world around us. This wouldn't be possible without that technology. So I think it creates new opportunities for human imagination, just in a different form.

# CHAPTER 5

# SHIMMERING ATOMS

Science without religion is lame. Religion without science is blind.
– ALBERT EINSTEIN

*KH: One of Einstein's papers that he presented in the first six months of 1905, his* annus mirabilis, *was a paper about Brownian motion.*

PC: Yes it was, and I had the fun of reading it again recently. It's a beautiful paper.

*KH: A beautiful paper?*

PC: He wrote so well. Apparently, in German he writes poetically, but I had to read an English translation. It is a beautiful paper. It was an astounding piece of work because it really ended the sterile debate that had taken place over a century about whether atoms existed. A lot of people had said 'There's no such thing as atoms', but the observation of Brownian motion and the explanation of it by Einstein laid that matter to rest.

*KH: It had been observed back in 1827 by the eponymous Robert Brown.*

PC: Yes, exactly. And he was a biologist. He was looking down his microscope at pollen grains suspended in water and he noticed this sort of restless, buffeting, jiggling motion going on with the pollen grains, and his initial reaction was that they might be alive, that he might be seeing the motion of living things. He was a smart guy and he thought 'Well, I'll go and try some mineral materials'. He even scraped a little bit of powder off the Sphinx in London, because that was very old and had to be very dead. And those particles did the same thing, so he thought 'They're not alive, there's something else going on there'. But he didn't know what it was. Indeed nobody knew what it was, and it was sort of a little mystery left there – all the way through the 19th century until 1905...

*KH:* *And this was one of the big questions that Einstein wrestled with: what is Brownian motion?*

**PC:** Yes it was, and he took a very bold step because, in a sense, the idea of this restless, buffeting, jiggling motion had been around for a while, but applied to atoms and molecules. And that was a theory developed by Ludwig Boltzmann who was trying to explain the behaviour of gases: how steam engines work and things like that – the laws of thermodynamics.

*KH:* *I knew we were going to get onto that...*

**PC:** And so Boltzmann said, 'If we postulate this idea of molecules wriggling around, we can explain the laws of thermodynamics.' That was terribly controversial; other scientists didn't like that at all.

What Einstein did was to say, 'Well suppose we use the same ideas that Boltzmann applied to molecules and apply it to a pollen grain.' This was a very bold idea. But it worked; it explained exactly the restless motion of the pollen grain, quantitatively – how fast it diffuses around. So that really showed that Boltzmann's idea of this thermal motion – this random, jiggling motion that is present in molecules – applies to everything. Absolutely everything.

*KH:* *So Einstein provided evidence of the existence of atoms by saying that atoms are what make the pollen grain jiggle?*

**PC:** Absolutely right. You see, the problem with the atom is that we couldn't see one. We can now, of course, but in those days nobody could see them.

*KH:* *But how did he prove that?*

**PC:** He proved this in a rather beautiful way, because with the pollen grain – which you could actually see in the microscope – it was possible to measure how fast it moved around over a certain time; how far it drifted in its motion. And from that number you could calculate what the size of the atom was – these atoms that were buffeting around it and pushing it from the side. You could actually work out the size of the atom. There had to be a theory of the atom to explain this buffeting motion of the pollen grain. In fact, what Robert Brown was seeing indirectly was the collective effect of all these little atoms and molecules around the pollen grain jiggling around it and pushing it. So with Einstein's explanation, based on Boltzmann's ideas, we could finally see, albeit indirectly, the effect of atoms themselves.

*KH: And did people read this paper and say, 'Don't be silly'?*

**PC:** Well, you know, just about everyone at that point was convinced and thought, 'This is real', but there was one 'hold-out' and this was Ernst Mach – after whom we call the speed of sound – 'Mach One' and so forth. Mach was absolutely opposed to atoms. He died in 1915 saying atoms don't exist. He was one who objected. But basically at that point the whole of physics was swept aside. And of course we subsequently had Rutherford and the discovery of the structure of the atom, so there was all sorts of new evidence that came about very soon after Einstein's paper to tell us that atoms really did exist.

*KH: And so Mach argued with the aforementioned Boltzmann about the existence of atoms?*

**PC:** Yes, Mach's point of view was that if you couldn't see it and measure it directly, it wasn't real ...

*KH: That's a good point!*

**PC:** Yes, Mach thought it was almost immoral to talk about atoms and molecules because you couldn't see these things. The really sad thing for Boltzmann was that, in the end, he took his own life. He took his own life perhaps because he was a person who suffered a lot from depression, but he was also much criticised by a large group of the scientific community led by Mach, who was a professor in his same university in Vienna. It was very hard for Boltzmann.

*KH: They talked together?*

**PC:** They did talk. They bumped into each other – they lived in the same city and they worked in the same university. So there was Boltzmann's greatest opponent right next door to him, and perhaps in the end he couldn't take it. And the tragedy is that he took his life in 1906 and wasn't aware of what Einstein had done.

*KH: Really?*

**PC:** Had he been aware of that paper...

*KH: It might have saved his life.*

**PC:** Well, he would have seen that he was completely vindicated, that all his ideas were right. So that's one curious historical twist.

*KH: On his tombstone, I read – Boltzmann's tombstone – is one of the central equations linking... I might have bitten off more than I can chew here. $S = k \log W$ linking the concepts of thermodynamics to the behaviour of atoms and molecules.*

**PC:** That's completely correct: $S = k \log W$.

*KH: Entropy equals constant times logarithm of the number of possible atomic arrangements.*

**PC:** Essentially, the probability. Boltzmann showed that you could explain all of thermodynamics by laws of probability.

*KH: Thermodynamics is simply... what?*

**PC:** Well, here you've got to take yourself back to the steam age. I mean, how do we make steam engines work really well? It's all about heat and work and pressure and those things. So the laws of thermodynamics came out of trying to understand steam engines. Take the first law of thermodynamics ...

*KH:* *Is heat a synonym for energy?*

**PC:** Heat is a form of energy. It's a form of energy that travels from something that is hotter to something that is colder. And you can convert it into work by a heat engine. That's what steam engines did. So then it was realised that heat is another form of energy. Consequently, the first law of thermodynamics is 'heat is a form of energy and the total amount of energy is always conserved'. You can't get energy for nothing. If you want to have work out of a steam engine, or a jet engine for that matter, you've got to put heat into it.

*KH:* *Because ... the amount of energy in the universe ...*

**PC:** Is fixed.

*KH:* *Is always the same ...*

**PC:** Well, that's a postulate of the first law of thermodynamics. The second law relates to 'how much heat can you convert to work?' Or ... (*dog barks in the background*)

*KH:* *This is my dog barking.*

**PC:** That's OK.

*KH:* *Clearly he dislikes the laws of thermodynamics.*

**PC:** He does. And actually he's a very wise dog because they're very tricky.

*KH:* *The night classes are paying off!*

**PC:** Yes, well, the second law of thermodynamics says that heat always goes from hotter to colder; it won't go from colder to hotter, even if energy's conserved. You can't make heat, of its own volition, go uphill, so to speak, in terms of temperature.

So thermodynamics is about laws based on things that you can measure. You can measure temperature and you can measure heat and work. But what Boltzmann postulated was that thermodynamics was all to do with the statistics of this random, jiggling, restless motion of atoms themselves. Of course, that idea was very, very controversial. Other physicists realised that if you based your ideas of thermodynamics on the motions of atoms, atoms which follow the laws of mechanics, then, because mechanical laws are reversible, so must be the second law of thermodynamics, a law we never, in practice, see reversed. So that was really at the heart of the principled objection to Boltzmann's work. But Boltzmann understood that while you could have the second law of thermodynamics violated, at least in a probabilistic sense, the probability was so small that you wouldn't actually see it in practice. It is ultimately by understanding this crazy, chaotic, random motion of atoms through the laws of probability that we can get an understanding of thermodynamics.

*KH: So what's entropy?*

**PC:** Entropy is a measure of the probability of a system. Systems tend to become more and more probable. For example, if you took a little bottle of ammonia and opened it in the corner of your room, ammonia molecules would come out and they'd gradually spread through the room and, after a period of time, you'd smell them everywhere in your room. The ammonia molecules will get themselves in a state that they explore all the possible places to be, so they maximise their probability. If they would all stay in one corner of the room, that would be a state of very low probability. So the spreading out of the molecules that makes the smell drift through the room is essentially the entropy getting larger. And it's the reason why all the gas molecules in our room don't suddenly rush to one corner, leaving us sitting in a vacuum and gasping for breath; they spread themselves around because that maximises the probability of their states. That's entropy in action. Entropy is relent-lessly increasing in the universe everywhere because of the way systems can explore all their possibilities and maximise.

*KH: Isn't the popular understanding of entropy that the energy deteriorates? That energy is lost over time?*

**PC:** Yes, the *quality* of energy deteriorates. The same amount of energy is always there, but the degree to which we can extract work out of it – get some value out of it – gets less with time. That's what we call the 'heat death' in the universe. But we don't have to worry about that on earth because we have this wonderful thing called 'the sun' pouring high-quality energy into us all the time, and that's what sustains life. But if you look at the entropy of the universe as a whole, it's in-creasing, so disorder is gradually taking over and that's an interesting fact of life.

*KH: So the process by which thermodynamics is useful when you're talking about atomic arrangements is that if you've got more heat, you have more busy atoms, and the busier the atoms, the less stable the material . . .*

**PC:** That's quite true, and every atom or every object has the same amount of thermal busy-ness, if you like. Take the case of a little dollar coin; if I put it down on a table the thermal energy is so small that it can't lift itself off the surface. If I take a pollen grain and put it on the table, it would have enough to sort of jiggle its own height. But if I took an air molecule, it would have enough thermal energy to get up to about 10 kilometres, which is why there's still air on the top of Mt Everest, even if it's a bit thinner.

*KH: So why doesn't all the air collapse to the floor, the way that if I dropped a dollar coin, it would drop to the floor?*

**PC:** Well, the answer is that the thermal energy isn't enough for the big object like the dollar coin to overcome gravity, but for the little air molecule, it's enough to let it get to the top of the atmosphere!

*KH: So this thermal energy means that nothing is really still, everything has this restless motion that depends on temperature, but we really only notice the effects directly for very small things like atoms?*

**PC:** Shimmering atoms. It's a lovely thought.

CHAPTER 6

# WHAT IS LIFE?

The highest wisdom has but one science – the science of the whole
– the science explaining the whole creation and man's place in it.
– LEO TOLSTOY

*KH: So, what is life?*

PC: You might well ask, 'What's a physicist doing talking about life?' I'm going to tell you some of my thoughts about life and, in doing so, I should apologise to the biological experts.

*KH: You mean because the division is that physicists only study 'not life'?*

PC: That's right and I think also there's a feeling, on the part of some physicists, that physics is the fundamental science of everything and that if you understand physics, you can understand everything that builds on it like chemistry and biochemistry and biology.

*KH: Yes, that's my thought – that if you boiled everything down to its smallest bit, you end up with physics or chemistry.*

PC: Well, that view was expressed by Pierre Laplace in the 18th century. He said that if we knew where every particle of the universe was and what its motion was at any instant in time, then in the future we could predict what was going to happen. And I think that's just completely untrue. In fact we know it's untrue by thinking about the most powerful computer that we could make in the universe to do such a calculation. Suppose we imagine a fantastic computer that used every atom in the universe to form its binary logic elements, its 'bits', and suppose we had to connect them all together by some communication, limited of course by the speed of light. We could work out what the biggest possible computer in the universe could do and, frankly, it couldn't do very much when it comes to biology. In fact, it couldn't even work out all the possible ways nature could make and assemble a single average-sized protein. So we know that this sort of reductionist, determinist view of the universe that physics presents at one level really can't explain biology.

*KH: No. This is, I suppose, the basis of the saying that for all the very clever scientists and computers and laboratories, nobody has yet been able to create a blade of grass.*

PC: Absolutely, and one blade of grass has such a mind-boggling complexity that physics and even chemistry can't come to grips with that. But biology comes to this in another way. Biology looks at nature in terms of laws of complexity; once you grasp this principle, for example the idea of natural selection that Darwin has explained, then suddenly there's an insight and you can see the way these complex systems work. But that doesn't mean that we should give up on the problem as physicists; we should try to understand if there's some aspect of the living process that we really can comprehend. For example, in the way I'm talking to you now, we could say that what is going on in this moment in my life is that my brain, with all its neurons and glial cells and axons and dendrites, is transmitting this information down through my nerve system.

*KH: Oh, don't think about it. You won't be able to do it if you think about it too hard!*

PC: Let me think. So ... those little neurotransmitters are communicating across the nerves, connecting down to the muscles in my mouth and my vocal cords, and all this activity concerns proteins in action, and that's, in one sense, life. Now, there is a sense in which we could use physics to describe life that way, and I think one of the things I'd like to do is to attempt to do that.

But to start at the beginning, if you were to say 'What is life?', I guess to most people it's a philosophical question. Why are we here? What are we doing? When we learned biology, we were told that life is about things that have movement, that feed, that excrete and that reproduce.

There is another sense in which we can see life as being just a multitude of organisms that exist in the world around us, 80 per cent of which are very tiny – the microbes. Most of life in its various forms is very unfamiliar, very strange and unicellular. So there is this fantastic diversity out there and the only thing that explains it is that they're all little survival machines for genes to replicate and that's what's driving all of life; it's those fragments of DNA – the genes – that survive, in the crossover process of sexual reproduction, for example, to determine behaviours – or phenotypes – that aid natural selection, whether you have long teeth if you're a carnivore, or whether your legs can run fast if you need to get away from carnivores. And so, it's the genes that are driving everything – it's the DNA.

*KH: So the criterion for whether something is alive or not is whether it has DNA?*

**PC:** Oh yes, indeed. It's the replicator molecule of DNA –

*KH: So nothing that doesn't have DNA is alive?*

**PC:** Well, nearly all, because some viruses carry their genetic code in a closely related replicator molecule, RNA.

*KH: I'm just wondering about some of those curious borderline spongy things.*

**PC:** Well, little bacteria, while they don't have nuclei, do have DNA, distributed around inside them.

So life began in the primordial soup, shall we say, maybe 4000 million years ago, with the formation of a replicator molecule.

*KH: It's great, isn't it? I love that expression – primordial soup.*

**PC:** Well, the origin is sometimes called the primordial twitch; the moment when life started was presumably when some molecule learnt to replicate itself and, once replication started, competition began in the soup for all the materials that made the molecules. Then the natural selection process came about. We've ended up with this very strange, complex array of multicellular organisms called plants and animals, including human beings. Most multicellular organisms have, of course, DNA wrapped up in a nucleus. But the way the DNA works – the way it makes that survival machine work in order to replicate – is via the mechanism of proteins, and the proteins are the molecules that really are the action molecules of life. And of course the DNA contains within it the code for making the protein molecule –

**KH:** *It's just an instructor, isn't it, DNA?*

**PC:** That's right, yes. You can say that DNA has two purposes essentially – one is to replicate and the other is to provide the code for making the proteins that will enable it to replicate. And the proteins do that by forming the organelles and the organs of life – the cells, the components of cells, the way cells come together and cooperate – and all this is being driven by natural selection, which produced this amazing diversity that we see now. But the protein is a most astounding thing because it has such enormous complexity – despite being made up of just 20 building blocks called amino acids.

**KH:** *Is that the definition of a protein? That it is made up of amino acids?*

**PC:** Yes, that's right. And the amino acids come together in a sequence that is determined by the DNA, so the DNA has in its coding sequence the message that is communicated to these little devices called the ribosomes where the proteins are made; the proteins follow that sequence as laid down by the genetic code in the DNA. When a protein is made, it simply becomes a long sequence of these building blocks. They might be a hundred or so building blocks, as in a small protein like haemoglobin, or a thousand building blocks in a large protein like the collagen in our muscles. But what these proteins all have is a special shape. And once they are made from the DNA instructions, they have to find their special shape. The protein works by having a particular shape and function and that shape is laid down in the code of the building blocks because certain parts of those blocks like to be together and certain parts are repelled from each other. But it's a huge task for that protein, once it's made, to find its proper shape – it's formed like a long piece of string and it's got to coil up into exactly the right shape predetermined by the code, and it's got to somehow find it.

*KH: Why is the shape so important?*

**PC:** The shape is so important because it determines the function of the protein. For example, if you think about a receptor protein that sits on the nerve cell that receives the little signal – this acetylcholine molecule that comes from the transmitter cell – the receptor protein has got to be just the right shape to receive the molecule that comes into its cleft and then causes the protein to twitch in some way so that a message is conducted through the cell –

*KH: Excuse me, this is a family programme! You mean it's a nut and a bolt thing?*

**PC:** Absolutely. So the shape is everything. But it's a massive problem in physics for a protein to fold to find its shape. It's like this problem of the biggest computer that we thought of – trying to work out how a protein would do that.

*KH: It's an odd thing, isn't it, to think that things as basic as proteins have to be exactly the right shape in order to do what they're supposed to do?*

**PC:** That's right and, of course, when there is a defect in the protein – when the DNA makes a mistake – that's the beginning of disease; when proteins have the wrong shape –

*KH: And they don't fit.*

**PC:** When they don't fit, we have malfunction and we have disease. But the process by which they find their final shape is really coming back to Brownian motion – this idea of restless, jiggling, wriggling motion

75

that goes on in all molecules because of the thermal energy. If you took a box of little matches where the matches were all jumbled up inside and shook that box of matches sideways a few times, backwards and forwards, they'd eventually line up all nicely – they'd find their shape. If you shook it too much, they'd all fly out of the box. So there's just the right amount of shaking to get them all to line up. And it's the shaking of the thermal energy that enables the protein that's waving its arms around and flapping about to gradually find the shape that it needs to be which has been laid down in those building blocks, and it needs just the right amount of thermal energy to do that. And, of course, that's why temperature is so important to life. We've developed a form of life that is based on the temperature in which we live.

*KH: So the same temperature enables all these different proteins to find their shape?*

**PC:** Yes, that's right.

*KH: They don't all need different temperatures?*

**PC:** No, most of life operates at similar temperatures. But what that means is that once we fix the temperature, proteins can't be too small or too big. Now, the reason they can't be too big is if they were, they'd never have enough thermal energy to get over those energy barriers – to get to their optimal shape. It would be too complicated. There'd be too many opportunities for them to get all tangled up like a fisherman's line when it's all got messed up on the reel. If it's too long, too big, it can't find its shape. And if the protein is too small, the thermal energy will be so big that when it's found its shape, the protein can shake out of that shape. So it needs to be just the right size. The curious thing about proteins is that they are all roughly the same range of sizes – maybe a hundred units of building blocks, up to a thousand units of building blocks – whether it's an elephant or a bacteria. All of life. We don't have big proteins in elephants and small proteins in bacteria.

*KH: No.*

**PC:** Proteins are proteins. They've developed to find their proper shape and that shape is determined by Brownian motion – it's determined by that Boltzmann energy that Einstein used to explain the random jiggling motion of pollen grains. So, in a sense, when Robert Brown looked down the microscope and looked at those pollen grains and thought 'I wonder if they're moving because they're alive' and he found that pieces of dust did the same, even ground-up dust from rock, he realised the motion wasn't life; it was just this strange phenomenon that Boltzmann and Einstein explained as thermal energy.

*KH: But it had an implication for life?*

**PC:** Absolutely. Ironically, when we look, we find that this 'inanimate motion' is driving life.

*KH: Did Einstein have any idea about that?*

**PC:** Oh, not at all, I don't think. And I think it's only something that has dawned on us in the age where we now understand life at a molecular level.

*KH: So do you think it's fair to say that Einstein was a genius? Is he revered because he found things that subsequently became important rather than things that were, per se, important at the time?*

**PC:** Of course, and when we look back in science, we can see the momentous discoveries that have led to so much greater understanding subsequently. Often at the time we don't know what an enormous step is being taken. We just have to think of Crick and Watson's discovery of the structure of DNA – very tentative, postulate, it's an idea – totally verified 10 years later but –

*KH: Let's run this up the flagpole.*

**PC:** Exactly, run this up the flagpole, and now what do we have? We have the genetic age. And so, you know it's only really once we understood

77

life at the molecular level that we could then start to see how the laws of physics were operating.

So the curious thing about this process – as I'm talking to you now the muscles are twitching away in my vocal cords, my brain is sending messages down through my nerves and these proteins are all action – is this: the proteins, having got their proper shape, are all on little excursions, in little wanderings around about their proper shape and that's the motion that's driving the life process; that's the motion that's allowing the nerve impulses, that's causing the muscles to contract. There is a lovely illustration of this idea when you look at muscles themselves, which are made up of proteins like collagens and myosins and kinesins, and the way that they work. What happens is that these proteins change shape, move and grab another part of the protein they're connected to and pull it along, change shape and move rather like a ratchet, and that changing of shape is caused by these Brownian motion excursions. And the direction in which they're driven is driven by the energy flows – this ATP (adenosine triphosphate) molecule that comes from the mitochondria that provides the energy that drives the muscles along. But without that Brownian motion, without that random motion, proteins couldn't do their job. So that's where the physics is: at any instant in life what's going on is a whole lot of wriggling and jiggling at the molecular level.

**KH:** *Is it useful to still have science divided into physics, chemistry and biology?*

**PC:** Well, it's sad in a way because it would be lovely to have a greater unity among physics and chemistry and biology. And I think that's the great challenge for physics and biology and chemistry – to find the overlap, as we struggle to understand the molecular basis of life. Biology is so far from physics in the way in which it operates and the challenge for physics is to try to deal with problems in complexity. That's something that physics is just nibbling at, at the moment. Complexity is going to be the great new challenge for physical science, I think, during this century.

**KH:** *So is chemistry not hot any more?*

**PC:** No, chemistry is very hot. In fact, chemistry's right at the centre of things. Chemists just have it made because, of course, all this complex motion at a fundamental level is about chemistry. After all, it's all about molecules, and chemistry is the science of molecules. So chemistry sits there as the bridge between physics and biology. Increasingly, we are seeing a greater unity between all these sciences.

# CHAPTER 7

# THE WAY NATURE WORKS

Thou cunning'st pattern of excelling nature.
– WILLIAM SHAKESPEARE

**KH:** *Professor Callaghan is about to address the question 'If the laws of physics are so simple, how come nature is so complex?' Or, as one of the developers of the theory called 'self-organised criticality' said, 'If the universe started with a few types of elementary particles at the Big Bang, how do we end up with economics and literature and life?' Is that a fair way of summarising what you're about to address?*

**PC:** I guess it is. But as well as that complexity, we just seem to live in a world that is governed by unpredictability, don't we? We have earthquakes causing a huge tidal wave off in Indonesia, we have new flu viruses suddenly appearing, and we have buckets of rain in the Bay of Plenty when it's normally sunny there . . .

**KH:** *And these people who worked out for themselves what they thought self-organised criticality was said if physics was so predictable, apart from your usual indeterminate bits where you can't tell where they are and how fast they're moving at the same time – Heisenberg's Uncertainty Principle – you've got a few basic laws. So how come nature is so apparently random and unpredictable?*

**PC:** Well, I think it's a feature of complex systems. Physics works very well at a deterministic level for certain sorts of problems describing atoms or molecules, but when you get large numbers of particles coming together, they behave in a more complex way. There do seem to be some underlying patterns there, and they're patterns that are a feature of the very nature of the complexity itself. Astonishingly, those same patterns appear in biology, they appear in social organisation, they appear in the way the stock market behaves. And, interestingly enough, it's physics that's been able to reveal those patterns.

And that's what this new theory of self-organised criticality is about. For example, the strange climatic events we see around the world that seem to be deviations from the norm, and other behaviours in the natural world around us, appear to be balance intermittently broken up by sudden, unpredictable changes. And then we go back to balance again.

You can see that in the stock market, looking at the prices, there are little fluctuations that occur day by day, and every now and then there are big changes, up or down. And of course we can't predict those, and anyone who could predict them would become extremely wealthy.

*KH: The model that one of the developers of self-organised criticality, Per Bak, uses is the sand heap.*

**PC:** The sand pile.

*KH: Is it useful to explain the sand pile to me?*

**PC:** Yes. I think the sand pile is a lovely example and it's the paradigm for the whole idea behind this.

If you take some sand and pour it out of your hand, a pile will gradually appear on the ground, and it grows as a sort of cone-shaped pile –

*KH: Quite happily ...*

**PC:** Quite happily. And as you pour more sand on, the pile gets bigger and bigger, maintaining a certain slope on the side. The slope is oriented at what is called the 'angle of repose'. Well, on the surface of the slope on the pile, every now and again, there are slips, or avalanches. Sometimes little avalanches will occur and sometimes there will be big avalanches, and often just dropping one more grain on top of the sand pile can cause grains of sand to go rolling down in an avalanche in some particular place.

*KH: So there seems to be no predictability or explanation as to why this big avalanche happens there and this little tiny one happens there?*

**PC:** Exactly. But what we do see in the pattern is that there are very few big avalanches and very many small ones. And of course we know that is the very same case with earthquakes, for example. Underneath Wellington, the earth's crust is buckling away as the Pacific Plate forces itself under the Australian Plate, and there are actually little earthquakes every day which we don't even know about, although geological sciences people measure them. We feel the bigger ones, of course, and every now and then there's a very big one. So the sand pile has exactly that sort of behaviour: a few big events and lots of little events. And the really common feature of the sand pile for all these complex systems is that the behaviour of each grain of sand in the pile really depends on all the others in the pile. You could just pull one grain of sand out or drop one grain of sand onto the pile and the whole thing would, for a moment, become unstable.

*KH: This has nothing to do with butterflies, does it?*

**PC:** Well, 'butterflies' is the idea of chaos – big effects from small changes in initial conditions. It's the idea that the flapping of a butterfly's wings in Tahiti can cause a tornado in Kansas.

*KH: This is not chaos theory.*

**PC:** It is not chaos – chaos is something a little different. Actually chaos, intriguingly, is not about complexity at all. You can get chaos with quite simple systems. You don't need to have many, many parts to the system to get chaos. I'll give you an example: suppose you have a pendulum – like the pendulum on a clock that's swinging backwards and forwards – and on the bottom of that pendulum you hang another pendulum, and on the bottom of that pendulum you hang another one again, so you've got three pendulums, all swinging, coupled together. That system will show chaos. It will show unpredictable behaviour so that any slight change in the initial conditions of the pendulum produces a very big difference in the outcome. That's a very simple system – just three pendulums. So you could make chaos out of quite simple systems; a few systems that are connected in some way in terms of their dynamics.

*KH: So not everything fits into self-organised criticality?*

PC: No, it's systems that are very big. I think ecosystems are exactly like that.

*KH: Just unpack the expression for me, 'self-organised criticality'. It's the criticality, or the critical state, where the avalanche happens . . .*

PC: Yes, exactly. The system has evolved itself to be on a point where it is critical – it looks like it's in balance but it's about to trigger sudden changes. In fact, if you look at the landscape of New Zealand, because of the uplift that occurs, we see these hills that have a certain slope on them, that angle of repose, and every now and again we get little landslips. That landslip could be triggered by a little bit of movement in the plate underneath, but it could be triggered by some cyclone that happens in the Pacific and comes rolling down to drop a lot of rain. So the behaviour of the land form in New Zealand is dependent not just on the ground underneath, but also on the whole climatic system. So it's very complex and, while it looks like it's stable, every now and again it becomes unstable. That's the criticality aspect. And natural systems that are very complex seem to evolve themselves to a state involving a very delicate balance that, every now and again, is interrupted by some catastrophic event.

*KH: Is the reason that scientists got so excited about this because it seems to be a theory that links the so-called 'hard sciences' with the 'soft sciences'? After all, hard scientists look at the nice, simple, physical, predictable way things work, and the soft scientists look at things like evolution. Is that why?*

PC: Well, I don't like to use the term 'soft science'. I think biology is just as challenging as physics. A better term is maybe 'emergent science' because in biology you explain events once you see them. In a sense you're looking at the historical behaviour and then you explain patterns. In physics you make predictions.

*KH: And the reason it's called 'emergent' and you can't make predictions is because it's assumed that it's so random?*

**PC:** Yes.

*KH: When we look at evolution, it does seem like a completely random thing that we are here today.*

**PC:** Well, there are too many parts to the system to explain it all in detail and I think it's really a great challenge for physics to try and understand that. Karl Popper once said that 'science is distinguished from pseudo-science by the ability to make predictions' and that's a huge challenge for science. We can't predict when the next earthquake will occur underneath Wellington. We know what's happening on the plate but we don't know exactly when the next earthquake's going to be.

*KH: So self-organised criticality doesn't help that?*

**PC:** It helps to a degree. For a start, we can see the patterns . . .

*KH: And the patterns are . . . that to an extraordinarily similar ratio, big things happen at a lower rate than small things.*

**PC:** Big things happen much less frequently than small things. That's right.

*KH: But in a specific ratio?*

**PC:** In a specific ratio, and this is the astonishing thing. There's a law of earthquakes, for example. If you plot the number of earthquakes versus the magnitude of the earthquake or, strictly, the logarithm of each of these, then there's a very simple linear relationship, called a 'power law'.

What it means is that an earthquake that's 10 times bigger is only one-tenth as likely to occur.

We find exactly the same law when we plot all sorts of seemingly unconnected events. For example ...

*KH: Traffic jams ...*

**PC:** Traffic jams, or, here's a beautiful one: if you take the frequency of words appearing in the English language, the most common word is 'the', the next most common word is 'of', the next 'and', and so on. If you plot them in rank order, you find that the tenth-ranked word is only one-tenth as likely to occur as the first-ranked word. And it's the same law. A man called Jack Levy did an analysis of wars that occurred between 1495 and 1973. There's 113 of them as it turns out, and in the number of big wars versus the number of small wars, you see exactly the same pattern. Big events are less likely to occur than small events, and it's the same scaling law.

*KH: I've been thinking about this. Are you sure it's nothing to do with the way we draw the graphs? Because it looks too tidy.*

**PC:** It looks very tidy. But there are some common features in all these systems, whether they are social systems, economic systems, or biological systems. First of all, as I said, they're very complex. Secondly, they're interdependent in that the behaviour of every part of the system is dependent on every other part. So it's a very non-local type of behaviour. In an economy, for example, what a few individuals can do can have some quite dramatic effects. And so this interdependency is something we can see in biological ecosystems; it's what we see in the earth's crust – because even if we knew exactly what was happening in the plate under Wellington, some event in Indonesia might trigger a slippage here.

*KH: All right. And that's why we can't predict that – because we might know the rough size of the next big one, but we won't know where it will be.*

**PC:** Well, we know where the energy is intensified in the plate movement under New Zealand – we know it's more likely in Wellington, for example, than Taranaki. But to know exactly when events will occur is very, very hard to tell because of this unpredictability.

*KH: OK, when. It's the when.*

**PC:** Exactly.

*KH: Have scientists managed to calculate this type of behaviour?*

**PC:** Yes. Some very nice computer models have been done now where you take a system and you make some interdependencies. A lovely one is, for example, on the fitness of species. If you take some set of biological species and you put some interdependencies in – like for every species there's a predator species and a prey species, so they're dependent upon each other, making a connection that builds right around all the members of the set – you can start to calculate. For example, you start to assign a degree of fitness to every individual species, and in each cycle of your calculation, you eliminate the least fit and add a new species to the set with a random value of fitness, watching how this whole thing evolves. What you find is that the entire ecosystem starts to evolve so that all the members get to a greater level of fitness and reach some sort of equilibrium, but every now and again there's some dramatic alteration when many, many species die out, and other ones appear.

*KH: Like a big earthquake?*

**PC:** Exactly, like a big earthquake. And if we look at the fossil record, we look at evolution and we see that's exactly what happens. Long periods of stasis or apparent equilibrium in the stability of species amid sudden extinctions.

*KH: This doesn't mean there is a God, then?*

**PC:** I don't think it says anything about whether there's a God or not. I suppose in a sense you could ask the question 'Who moves that one grain of sand in the sand pile to cause the whole thing to avalanche?', and of course that's the thing we can't predict.

**KH:** *Who drew that divine straight line in the sky?*

**PC:** And who made the pattern underneath?

**KH:** *That's for another conversation perhaps.*

**PC:** Another conversation. And meanwhile, the option's open for those people who want to believe in God, I'm sure.

## CHAPTER 8

# EVOLUTION

Man with all his noble qualities . . . still bears in his bodily frame
the indelible stamp of his lowly origin.
– CHARLES DARWIN

*KH: Let's talk about the theory of evolution. I'm saying 'the theory of evolution' because of course in America there is an increasing number of people in positions of power in the education system who are saying 'Remember, it's only a theory, and we have to do the "intelligent design" concept alongside it in order to give our children a balanced education.' To which you say . . .*

**PC:** Well, I guess Newton's laws are a theory; quantum mechanics is a theory. But I'd have to say, as a physicist looking at the theory of evolution, that it's hard to find a more successful and robust theory in the whole of science.

*KH: How do you mean?*

**PC:** Well, since Darwin proposed the idea in 1859, there were so many opportunities for it to be proven wrong. It was there to be falsified, with all the subsequent discoveries in physics and earth science about the age of the earth; with the new discoveries about genetics. But to the contrary, every new discovery we make only seems to reinforce the idea of evolution and natural selection as the way in which life has come about.

*KH: Wouldn't the creationists say 'That's because you're making the rules' or 'You're wilfully not looking at the big gaps'? Wouldn't they say, for example, 'Where is the fossil evidence for the transitional . . . bits?'*

**PC:** There are transitional forms. For example, there's evidence in the fossil record of intermediate forms like bird-reptiles. But yes, of course there are gaps in the fossil record, and there are some life-forms that didn't lend themselves to fossilisation at certain times. But the fossil record is only just one part of the evidence for evolution.

*KH: Is it? What's the other evidence?*

**PC:** Well, for one thing we can see evolution and natural selection working on a day-by-day basis: we know that when we try to enhance it in an artificial way, with breeding, that we can produce really remarkable divergences of life-forms in, for example, the variations of dogs. We know that bacteria evolve and develop resistance to antibiotics. We see animal species showing the result of evolving through natural selection. A remarkable example is the African elephant. There's a very small probability that an elephant can be born without tusks, but with all the poaching of elephants, that probability has apparently increased to about 10 per cent of the African elephant population. Because if you have the gene for 'no tusk', you've got a better chance of survival.

*KH: Lamarck presumably would've said if you take enough tusks from enough elephants then eventually elephants will lose all their tusks – but it's really the other way around, right?*

**PC:** It's the other way around, yes. That's the strange way in which it works. But getting back to that independent evidence for evolution, there is evidence now through the whole of our understanding of molecular genetics, an understanding based on a unity of life seen through the way in which DNA, as the basic replicator, is present in all life-forms. And I suppose some of the most powerful evidence is now seen through the whole tree structure related to evolution – the way in which the taxonomy of life works. We break life into the various domains and kingdoms and phyla and families and so on, the idea of Linnaeus, the categorisation of life. But of course we can do the same thing now with molecules as well. With DNA we can see the same kind of tree structure there. And there is such a lovely expression to describe this, which I just have to say . . .

*KH: Control yourself!*

**PC:** It's called 'consilience of independent phylogenies'. What that means is that when you look at the mathematical relationships in the tree structure of DNA they almost perfectly match the same relationships that are in the structure of the categorisation of life. There's wonderful mathematics behind that, and in fact we have some of the leading people in the world in that area working in New Zealand.

*KH: I'm not sure you've explained the consilience of independent phylogenies...*

**PC:** Oh, I'm so sorry about that. Consilience simply means that there is a similarity in the mathematical relationships of the tree structure. If you look at the patterns of the relationships between DNA in different species it follows almost exactly the patterns that are present in the classification of life-forms.

*KH: That doesn't seem very scientific, somehow. It seems a kind of romantic and artistic view of it but not very scientific. Like, you make a certain pattern and it's easy to find another pattern to fit it if you jiggle it around a bit.*

**PC:** Well, there are a couple of things you can look for: you can look for something called homology. For example, in whales, with their flippers, you find the same sort of bone structure as you find in the limbs of land-walking mammals.

*KH: Right...*

**PC:** So that's one example of something you'd expect from descent from a common ancestor of some sort. And remarkably you see the same sort of thing in molecules: there are proteins like lysozyme and alpha-lactalbumin that do totally different things. I mean, lysozyme protects us from bacteria by attacking their walls and alpha-lactalbumin helps with our growth and the development of our immune systems, yet the protein structures are almost the same. So they've come from some common background where they've deviated through mutation in some way and have ended up doing completely different things. There are all sorts of things about evolution that just lock into place. And it's very interesting that evolution has by no means resulted in a perfect solution in adaptation. It is so obvious that we sort of 'make do with what we've got'. It's not a very 'intelligent design' in some ways when you look at various life-forms, and that in itself lends some support to our ideas about the complex pathways of evolution.

**KH:** *Who was it that said 'If the intelligent designer made us then we'd be breaking his windows'?*

**PC:** Yes, that's right. It was a comment in a recent *Time* magazine by a scientist in the United States. There are some curious things about humans, for example. We have these inverted retinas. Not all eyes have developed with inverted retinas, but ours have. The disadvantage of that is that we've got to take nerves out and bring them to a certain point on the retina . . .

**KH:** *Which gives us a blind point . . . spot.*

**PC:** Which gives us a blind spot. Exactly.

**KH:** *Let's talk about that eye because this is the basis of a lot of creationist arguments or intelligent design arguments. Namely, how could the eye, in all its complexity, have possibly evolved, given that they've got this thing called 'irreducible complexity'?*

**PC:** Yes, I think that really goes to the crux.

**KH:** *So you take one little bit out and the whole damn thing falls apart. So how could evolution possibly be responsible for this?*

**PC:** Well, I think that understanding this, understanding how the eye could evolve, is probably the most beautiful and powerful insight that one can get through understanding evolution. Because the greatest misunderstanding about evolution is that somehow things happen just by chance. How could the eye, with all of its features, have come together by chance? But of course evolution doesn't work that way at all. Evolution is a slow process of gradual change with natural selection kicking in to favour those changes that are advantageous, and to severely and brutally punish those changes that are not.

*KH: So chance is not an adequate description...*

**PC:** Not at all. And, in truth, the eye appears in various forms of life and has evolved quite independently – in about 40 different forms. The way the human eye works is totally different to the way insect eyes work, with their compound nature. Once you've understood how evolution works, it's quite easy to see how the eye would have developed. In the early forms of life, bathed in the light coming from the sun, there would have been a real advantage for any organism that had any sensitivity at all to the presence of light. Starfish, for example, have cells that are sensitive to the mere fact that light is there. But then imagine that you wanted to know where the light is coming from. If your little light-sensitive cell was moveable, you could – like maggots do now – move the cell from left to right and see which way light is coming from. But it would be far better if you had more cells and if these formed some sort of a concave structure. So you can go through and see how every little change that caused the organism to have a slightly more effective vision would have given it a huge advantage. And so the driving force behind natural selection for the evolution of the eye is so strong that it's not at all remarkable to envision how it could have come about. As Richard Dawkins says in *Climbing Mount Improbable*, it's not a matter of evolution attempting the impossible climb of an enormous cliff to achieve the end result of the eye. That's not the way evolution works. Evolution goes through the back side of the mountain on very gentle slopes, working slowly, step by step, over hundreds of millions of years.

*KH: Nevertheless, there is an emotional incredulity, that one can understand, attached to how this could possibly come about without somebody planning it.*

**PC:** There is. It seems so counterintuitive. But that is, of course, the way science is. Science is a means of discovering knowledge that defies common sense. I mean, the fact that the earth goes around the sun and not the sun around the earth defies common sense. If you look out into the sky you see the sun going around the earth. So science actually enables us to understand things that don't accord with our common sense but are true at a deeper level. So, of course, we who live our short lives compared with the whole span of life on earth have a view, which is our common sense, that lets us down. The way science works is by applying observations and requirements of consistency and falsifiability. And through an openness of discussion and debate, truth emerges.

**KH:** *Some people try to have their cake and eat it too, of course, and say that we can believe in evolution as well as an intelligent designer because the intelligent designer started the evolutionary process. Do you think that that is a bit of a fence-sit?*

**PC:** I think that the people who propose intelligent design go further than that. They actually propose that the designer is kicking in at every point of the evolution of a species. But if someone says someone started off the universe at the beginning and evolution all happened after that, then I guess that's a reasonable form of belief. It's not one that I happen to subscribe to. But I don't see any incompatibility between religion and science.

**KH:** *Unless you are a literal interpreter of the Bible . . .*

**PC:** Yes, and you can be a literalist to the extreme that you believe that God created the earth in six days, 6000 years ago. But of course the evidence doesn't support that, and to what degree are we prepared to simply ignore the evidence and just set our minds against the way in which science requires evidence and requires consistency?

**KH:** *How do you feel, then, about the moves in the United States among a number of states to challenge the laws of evolution? The argument seems to be that yes, we'll teach evolution in schools but we'll also teach them other ideas so that we educate our children roundly.*

**PC:** I think it's quite appropriate to talk about other creation ideas in sociology and in history, but science is not about setting the non-scientific view of the world against something we do understand as the way science works. I think it's very dangerous when politicians or people with a religious perspective want to influence the way in which science works and the way science is taught.

*KH: But the message from that, Paul, is that lay people should stay out of science, and I think you'd argue for people to get more involved and more knowledgeable about it.*

**PC:** Yes, absolutely.

*KH: You can't get more knowledgeable about it unless you get a counter-argument.*

**PC:** No. The problem with 'intelligent design' versus evolution is that intelligent design doesn't use the paradigms of science. It doesn't open itself up to scrutiny, it doesn't subject itself to peer review, to publication, to the normal debates and normal methodologies of science. It comes in with a fixed view of the way it is, requiring the existence of a creator, and therefore raising new, unanswerable questions as to what the creator is. It is based neither on observation, nor on consistency with other science knowledge, nor on science methodology. I think that's what makes it fundamentally unacceptable to scientists. Scientists do accept that many different theories can compete for being successful, and evolution is one theory – it's true. But it's the only theory that seems to work in a consistent way. And it has a remarkable beauty and power in explaining all the complexity and diversity and, I must say, perfidiousness, that exists in life.

*KH: Perfidiousness – why?*

**PC:** Well, just think of the diseases that cause the most terrible suffering among human beings – what intelligent designer would've made that?

99

*KH: Right.*

PC: We *would* throw rocks through his or her window, wouldn't we? So it's a strange and brutal and cruel nature out there – as well as being a beautiful world – and of course natural selection explains that very well.

*KH: What would be one thing that would make you rethink the validity of evolution?*

PC: Oh yes, there are some falsifiability tests which are quite clear. The biologist J.B.S. Haldane once suggested the fossil of a rabbit from Precambrian times. I mean, finding that would just blow evolution out of the water. And of course had Darwin, with his predictions, been followed by a discovery that the earth was only a few thousand years old, that would have blown evolution out of the water. Of course, we now know that the earth is four-and-a-half billion years old. So there are many things that could have destroyed evolution as a theory, and still could – for example, a fossil that was totally inconsistent – but it hasn't happened. That falsifiability is what makes evolution a proper scientific theory.

*KH: The suspicion is, of course, that people are only looking for what fits into the theory.*

PC: Well, I think that any scientist who found a rabbit fossil from Precambrian times would be very famous and could well win a Nobel prize. So there are lots of incentives for someone to do that.

*KH: Keep looking for those rabbit fossils?*

# CHAPTER 9

# SEX

Sex is something I really don't understand too hot.
— J.D. SALINGER

*KH: We're talking sex and here's Professor Paul Callaghan, looking dismayed. Actually, we are going to talk about sexual reproduction. Here's the question, then: why did it happen? Presumably at one point everything was asexual reproduction.*

**PC:** That's right, and there are still many examples of asexual reproduction.

*KH: So how come we have sexual reproduction? Matt Ridley points out in his book* The Red Queen *that if we're into reproducing as much as we can, sexual reproduction is very inefficient because you've got to find a person of the opposite sex to do it with, whereas in asexual reproduction you can do it yourself.*

**PC:** That's right. There are all the courtship rituals which waste time –

*KH: Waste of time, Paul!*

**PC:** There's the fact that you could produce males, and males don't produce offspring directly – completely inefficient . . .

*KH: Waste of time, Paul!*

**PC:** So why do it?

*KH: So why do it!*

**PC:** The simple answer is that sexual reproduction allows for the accumulation in one individual of beneficial genes that have been mutated or formed in various previous individuals. And we sure need that, because life is a struggle. You only have to look at the Black Death, which wiped out half the population of Europe in the 14th century, or the Great Plague in the 17th century. Some people had the genes that enabled them to survive; those genes were accidentally created out of mutations. And it's the shuffling process of genes that happens in meiosis – when the sperm is made and when the female eggs are made. That's the point when our mother's and our father's genes are shuffled up just to make that bare set of 23 chromosomes that are in the egg or the sperm. That's when the opportunity comes to build up those mutations.

*KH: I think Matt Ridley would say that that's a consequence of sexual reproduction and not a cause.*

**PC:** But of course, and it's a very complex matter looking at the cause, because genes don't have any control over the future. But if, in the various copying errors or the various mutations that occur, something happens that is beneficial, then natural selection would favour that. So looking backwards, you'd say 'How does this marvellous set of events occur?' or 'How does this organism result that seems to have all these advantages?' We forget all the copying errors or mutations that led to disadvantageous organisms – they just got wiped out in history. So what we see surviving looks to be a wonderful machine – a human or any other form of life – but of course we don't see all the failures that have occurred through evolution.

*KH: So who are the asexual reproducers?*

**PC:** There are many organisms, especially at the single-cell level, but there are examples of higher life-forms, like the whip-tailed lizard, which has only females that produce females. But with all asexual repro-duction there's a real danger because if you're not shuffling those genes you don't necessarily have the possibility of accumulating beneficial mutations – for example, the way we humans do in defending ourselves in the battle against the parasites.

**KH:** *How does that work when the females produce females? Where do you get the males from?*

**PC:** The female produces the egg which hatches a female.

**KH:** *There are no such things as male whip-tailed lizards?*

**PC:** Not at all. And it raises a very interesting question of what is 'maleness' and what is 'femaleness'.

**KH:** *How do you mean?*

**PC:** Well, if you accept the idea that it is advantageous for cells to swap their DNA, to produce the shuffling process that goes on, then you don't actually have to have a male and a female to do that. There are many fungi – for example, isogametes – that just shuffle their DNA between them.

**KH:** *Right. So why maleness and femaleness?*

**PC:** Well, there's a very interesting thing. If you imagine a situation where eggs could shuffle their DNA, imagine that one egg happens to be a little bigger and another egg a little smaller. Then the smaller egg has a lesser food supply and it would be a huge advantage for the small egg to find a big one to join with. That would give the small egg a bigger food supply. The small egg wouldn't bother seeking out the medium-sized ones; it would go for the big ones.

**KH:** *Right... so...?*

**PC:** So that starts to drive a strategy where the big ones get more and more favoured and the small ones become more and more mobile and take advantage of the big ones, and the whole thing splits apart. And so eventually you have a division into very small eggs that can move around very rapidly and very large ones that become their targets.

*KH: Is there any evidence that this was what happened – that the sperm evolved from a small egg?*

**PC:** That's the pattern that seems to be everywhere in nature. And what defines maleness is multiplicity of the gametes, mobility and smallness. That's true with plants as well – the pollen is very small; it's mobile. The female egg or the female part of the plant is immobile. If you look at a frog, for example, there's no genitalia to say if it is a male or a female. So what defines maleness in all species and what defines femaleness? It is that females produce few eggs; they are very large; they are immobile. Males produce very small gametes – sperm, as we call them in mammals – which are extraordinarily numerous and very mobile.

Remarkably, that separation starts to drive the behaviour of the species as well, because if you're a male producing vast numbers of these gametes, you can have multiple partners and there's no limit to the number of offspring a male can have by mating with many females. But a female only has a few eggs and a gestation period, so she's limited in terms of the number of offspring that are possible.

*KH: Biology is destiny, then, is it Paul?*

**PC:** It does seem that way!

*KH: It's pretty inescapable, isn't it?*

**PC:** It is inescapable. It drives behaviours, and it means that in the case of the females of a species there's an advantage to having faithful males who hang around and help bring up the offspring. There's also an advantage for the females being a little coy and sounding out the males beforehand at the cost of not being able to choose a partner so quickly.

*KH: Isn't there also an advantage – and I think there is a species of monkey that has proven this – for the women to have as many male partners as possible so they all think that the baby is theirs? Lest I sit on the moral high ground!*

**PC:** Exactly. You can find every single example. You can find promiscuity in nature, you can find polygamy, you can find monogamy – every species has developed its own strategies around this, and these are what are called environmentally stable strategies that work for that species.

*KH: So how is maleness and femaleness determined?*

**PC:** Well, we humans have got a very strange way of creating maleness: we do it through the Y chromosome –

*KH: Which is a lottery . . .*

**PC:** It's a sort of lottery and it's a very interesting story, because not every species does this. I mean there are certain fish, for example butterfly fish, in which female fish can turn into males by a social signal when there's an absence of a male around. Or there are turtles and alligators that determine the maleness or the femaleness of eggs by the temperature at which they hatch.

*KH: Bees and wasps as well.*

**PC:** Exactly. But we've got this strange thing, the Y chromosome, which is only carried by the male –

*KH: They'd think we were bizarre holding the baby up in the delivery room, saying 'It's a boy!' or 'It's a girl!'; they'd say, 'Why don't they know already?' And of course we do sometimes.*

**PC:** We do. We know from the genetic information.

*KH: But we can't control it other than through very controversial means . . .*

**PC:** That's right. But then, of course, part of human reproduction is asexual. In every cell in our bodies there are the mitochondria. Now, mitochondria have DNA that is passed on only from the mother. It only goes down the female line. They are probably remnants of ancient bacteria that invaded or got incorporated in the cells. They consume oxygen to produce energy, so they are wonderful little things, producing the energy that we need in the cells. And of course in the female egg in the human species there are thousands of mitochondria providing the energy – none at all in the sperm – which is why those mitochondria reproduce down the female line only, just as the Y chromosome follows down the male line only. It's through the history of the mitochondria that we can tell where our common maternal ancestor is. It's through that that Allan Wilson, a New Zealander working at Berkeley, was able to tell that we all came from a common female ancestor about 150,000 years ago – probably somewhere in Africa.

*KH: Yeah, but that's disputed, isn't it?*

**PC:** Well, it's a statistical common ancestor. I mean, there's not necessarily one individual person – there's a sense in which there's a convergence of the DNA.

*KH: Why is it specially designed through evolution that we should have a 50/50 breakdown of the genders?*

**PC:** Oh, it's obvious when you think why it has to be that way. Suppose, for example, there was a gene that tended to produce more daughters. That wouldn't necessarily be a problem because it only takes a few males to produce offspring in many females. So having lots of daughters, you could be sure that they could all find a male to produce offspring. But then those few males who are left would have an enormous opportunity to widely spread their genes among the many excess females available.

*KH: Which would be good for the species because then you could produce zillions . . .*

**PC:** It would, but it would be very good, then, for a gene to produce males. So suddenly the advantage would switch over to a gene that tended to favour producing sons, because the few males around who had this excess of females could produce many, many offspring. This makes the gene highly effective.

*KH: But they wouldn't only have males, though?*

PC: No, but if there was a gene that favoured producing males and there was a preponderance of females, then those males would have a great advantage in spreading their genes. Do you see the point?

*KH: Yes, I do.*

PC: So then suddenly the advantage would switch back to having sons. And so nature does this and the genes get into balance. There's a battle of the sexes and the natural outcome from that is a 50/50.

*KH: The way Matt Ridley explains it – and it took me a while to get this – is that if one segment of the population is having sons, it's rewarding for the other segment to have daughters, and if you stray from the 50/50, it'll pay to have more of the rarer gender. So it's like supply and demand.*

PC: Exactly, and we have to think of this from the point of view of the genes. It doesn't matter about individual gene-carrying organisms – like humans – it's the genes themselves that count. Overall there has to be a 50/50 balance.

*KH: So evolution has built in an unconscious mechanism for altering gender?*

PC: Exactly, although in humans we know that the male may be in trouble because the Y chromosome is suffering successive damage, generation upon generation. It suffers this damage because it resides in the testes of the male where the mitochondria are very active. There's a lot of cell division taking place there – much more so than in the female egg – so the copying errors start to accumulate. In the female egg the mitochondria are switched off and all the free radicals that are resulting from the energy production are not available to actuate damage. So the Y chromosome is a bit of a genetic mess and, successively, male fertility is dropping. Maybe, in about 100,000 years, male fertility will be down so low that there will be a real problem for the human species.

*KH: So not only no more sons, but no more daughters either.*

PC: Well, there's got to be a solution found somewhere. Maybe the gene that switches on maleness might need to jump to the X chromosome or somewhere else.

*KH: When you say there's got to be a solution found somewhere, you mean evolution will design a solution?*

PC: Or human beings will become extinct.

*KH: Well, that's right, but has evolution got a chance to find that solution?*

PC: Yes, because if a mutation manages to come about that avoids this problem, then that line will suddenly have an advantage and that genetic outcome will be very advantageous.

*KH: So, given the selfish gene, what about altruism, then? Do you think it exists?*

PC: Well, the gene may be selfish but that doesn't mean that we as individuals have to be selfish, that's for sure.

*KH: Sit on the fence, why don't you!*

PC: Indeed, and in fact there are lots of advantages in being altruistic in protecting our own genes, obviously of our kin, but also of our fellow species. Altruism is actually a feature you can understand in terms of the selfish gene.

*KH: My understanding of altruism is that it's only altruism if you are prepared to do it even if it may bring you harm.*

**PC:** That's right. Even if it may bring you death. But, for example, it pays your genes if you die saving four or five of your siblings; your genes are going to do a lot better in the long run.

**KH:** *Yes, but what about four or five horrible people that wanted to move into your house and kill your pet?*

**PC:** It's most unlikely that your instincts would make you want to save them.

**KH:** *We'll talk more about that.*

CHAPTER 10

# THE GENE, ITS SELFISHNESS AND ALTRUISM

Know then thyself, presume not God to scan,
The proper study of Mankind is Man.
– ALEXANDER POPE

*KH: Evolutionary scientist Richard Dawkins popularised the expression the 'selfish gene', meaning the actions of organisms can be explained by their programmed aim of perpetuating their DNA. So, can the selfish gene explain what we call 'altruism'?*

**PC:** Well, the genes can often achieve their selfish ends through paradoxically unselfish behaviour on the part of their 'survival machines': the organisms. It's very important in understanding evolution to distinguish between what the genes are up to and what the survival machines or, to put this in a human context, what people are up to.

*KH: But we are our genes...*

**PC:** We are a product of our genes in the sense that our genes laid down some 'software', software that guides the way we behave. That software sets us up with certain predispositions. For example, a child's cry – pain tells them don't do it again. A child smiling – a pleasant feeling, a nice sweet taste in the mouth, say, is good, so let's do that again. So that software laid down for us is what guides our behaviour. And it has been laid down through a whole series of evolutionary battles, involving the competitive survival of organisms, the successful survivors having the beneficial mutations. Of course, all the unsuccessful mutations that led to behaviours that were not successful lost out in the struggle. Their genes died out.

*KH: So if we see unselfish behaviour as a property of many living organisms, could that be explained in terms of the genes getting their own way as well? In other words, can unselfish behaviour be seen as, effectively, selfish behaviour?*

**PC:** An obvious example would be the case of kin altruism where, for example, if a person were to give up their life saving several of their siblings, actually their genes would win out. The organism might die but the genes would survive. So if people have been programmed to have a certain amount of unselfishness and altruism towards members of their own family, that's actually quite advantageous for the genes.

**KH:** *Where does self-determination figure in all this?*

**PC:** Well, of course we do have the capacity to overcome our selfish genes. We as individuals – particularly as humans – are uniquely placed to be altruistic and overcome the programmes that are laid down within us. But even with animals we see examples of paradoxically unselfish behaviour. An example would be in a flock of birds when a predator appears. One of the birds may suddenly squawk. This would presumably attract attention to that individual bird and you might wonder why it would take the risk. But it may be that by alerting all the others, the whole mass of birds can fly to a tree and stay together – to have the safety of numbers. So what appears to be an unselfish act of bravery on the part of one bird might help the survival of all the birds, and actually help the survival of the genes in that particular bird who made the alarm.

**KH:** *And in the same way bees commit suicide when they sting to protect the hive.*

**PC:** Bees are a very interesting example because the degree of gene-relatedness among the worker bees is much higher than gene-relatedness among normal siblings. But, of course, the worker bees are sterile, they have no capacity to reproduce, so therefore their genes do very well out of their unselfishness. And you can think about a hive of bees in the same way as you can think about all the cells in an organism. For example, there are many cells in our bodies that are dying all the time, but the important cells that determine our reproduction are our gametes – our eggs and our sperm. The rest of the cells don't matter at all in terms of gene propagation, except to provide a supporting organism. You can look at a hive of bees in the same way – certain bees have a role that's crucial for the reproduction, but others are just part of the whole organism.

*KH: You could argue, then, that what makes us human is our ability to overcome our selfish genes.*

**PC:** I think that's very true, and of course as humans we've developed capacities to do that through using language. Language is a mechanism whereby we can transmit ideas, and ideas and cultural values become a little bit like genes in the sense that they themselves can replicate – they have to be passed on; they have copying errors; there are issues of fidelity of the persistence of the ideas.

*KH: What are you thinking of there?*

**PC:** Well, I'm thinking of something that Richard Dawkins has called 'memes'. I guess religious beliefs would be an example of a cultural value that has been passed on and is quite persistent – it survives over many, many generations. Even in simple little things like popular sayings that penetrate our minds, that get passed from one individual to another and become part of the popular culture of a time ... perhaps even musical phrases.

*KH: But we're not born with that knowledge; we have to learn it.*

**PC:** We have to learn it, yes, and there are some examples of that in animals too. I think the New Zealand saddleback exhibits many different songs ...

*KH: In fact Dawkins talks about that, doesn't he?*

**PC:** That's right. Birds learn particular songs which are passed on. So not all of the behaviours that we have come about from our genes. Clearly there is as much that is nurture as there is nature, and the idea of the meme as a replicating entity that determines our behaviour is an important part of that idea.

117

**KH:** *In fact, we've heard from Ingrid Visser who studies the culture of orcas off the coast of Argentina that they learn to hop up on the beach and seize baby sea lions. Whereas if you confront an orca in New Zealand with a beach and a baby sea lion, they supposedly wouldn't know what to do with it, but they would learn if they were put off the coast of Argentina. I am wondering, then, are 'memes' just Dawkins' way of getting around genetic determinism?*

**PC:** Yes, I think you have to accept that genetic determinism is not everything, but it explains so much. But, coming back to altruism, I suspect we find that a degree of altruism is confined to the species and doesn't go outside it. You'll see that throughout nature, and part of the reason is that there's a *symmetry* within a species. But where there's a great *difference* in the capability of an organism, it's unlikely that you'll find altruism between those different species. Within a particular species life is not a zero-sum game: it's not like a game of football where someone has to win or someone has to lose. In life it's possible for everyone, within a species, to be a winner.

**KH:** *Everyone?*

**PC:** Well, let me explain it by example. If there's a competition over food supply or over the right to mate with a female, then a conflict situation arises. So how do you deal with that? Do you fight to the death, or do you deal with that in a slightly more gentle way where you don't fight too hard and you give up when things get a little bit rough? There's a dove strategy and there's a hawk strategy. There are wins and losses if you look at this from either way, but the loss is far the greatest if you die in a fight. So to be a hawk can be extremely dangerous, and if all members of a species behave like hawks, some of them are going to get seriously hurt and the losses can be very severe. It pays them all to behave like doves in a sense. But the problem then is, if a hawk comes along, it invades that strategy. Suddenly a hawk among a whole lot of doves starts to win big-time. Eventually there tends to be some sort of a balance, something that is called an 'environmentally stable strategy'. What emerges in a species is a bit of hawk-like behaviour and a bit of dove-like behaviour. Even in each individual member of the species you'll see elements of those sorts of behaviours. But that altruistic streak, the dove-like streak, plays a role because it leads to a stable way for everyone to win and for the genes to propagate effectively.

*KH: It all sounds very mechanical.*

PC: It does sound very mechanical, doesn't it? And of course I don't wish to imply that human behaviour is entirely determined that way. As I said, cultural values, ethical values, are very important, but that software that we are born with – the nature part of us, the part that we are in a sense having to overcome – is there as a platform on which we develop as human beings.

*KH: Let's just go back to Dawkins for a moment. He gives the example of contraception to show how we've overcome our genetic programming.*

PC: Yes, it's a classic example, of course –

*KH: But couldn't that simply be seen as another example of the selfish gene at work? If you limit your number of offspring, then the offspring you do have will have a better chance at survival.*

PC: Yes, of course. It changes the whole pattern of the evolutionary process – every intervention like that –

*KH: So is contraception an example of the selfish gene or is it an example of the meme?*

PC: I think it's an example of the meme interacting with the selfish gene –

*KH: Oh, sit on the fence again then!*

PC: I am sitting on the fence, aren't I? I guess what I mean is that life is far more complex than the selfish gene alone. In fact, I think that is what is so interesting about human behaviour and the role of memes, namely the very important aspect of cultural values in our lives and the way in which that determines how we behave and how our species develops.

**KH:** *So you could also say that religion is either a way of overcoming the selfish gene, or another manifestation of the selfish gene in terms of holding communities together.*

**PC:** Well, I'd like to express a personal view on this: I don't think that religion makes people any more altruistic or any more unselfish. I suspect that you will find just as much selfishness and lustfulness and wickedness, if you like, in people who have a religious perspective as in people who are unbelievers.

And I think that these ideas of ethical values and moral behaviour are something almost independent of religious belief, something incredibly important to humans in terms of their cultural values. We know that it is very important for most species to have an element of altruism in order to survive, so something deeply embedded in our natures is that we need to look at our behaviours in a way that is considerate of others. So I don't think religion necessarily does that for you. I don't believe that religion necessarily makes people ethical in their behaviours or altruistic, but I also don't believe that science does either. Science tells us nothing whatsoever about whether we should do something or what is the right thing to do; it's merely a way of attaining knowledge about the world. That's a struggle we just have to have with ourselves as individuals: how we behave ethically.

**KH:** *You're suggesting that they're mutually exclusive, science and religion?*

**PC:** I don't think that they're mutually exclusive, but I think that in science it's very hard to have a religious perspective. There's a sense in which I think that as a scientist one can also feel that the God of the church just isn't big enough. What I mean is that when you look at all of the universe – all that we see in the vast aeons – through evolution – and you see that humanity is less than a million years out of 15 billion years of the universe, it's very hard to regard the universe as a platform or a stage on which God can watch the playing out of good and evil among human beings. It's hard to believe that

THE GENE, ITS SELFISHNESS AND ALTRUISM

human beings are so central to the whole of the universe when you live in the world of science. And the other thing about a scientist is that you are taught to doubt. Doubt is at the very heart of science.

*KH: And at the heart of religion?*

**PC:** At the heart of religion is *faith*. Faith is partly about accepting a belief on authority. And in science one never does that. No rule is unable to be tested, and the ultimate judge in science, of course, is observation. Observation is the cruellest judge of all. And in science we require rules to be consistent, but they have to be testable; they have to ultimately withstand the rigours of observation.

*KH: Do you think we're genetically programmed to search for some sort of religious belief?*

**PC:** I think that's quite possible. I think that it's possible if we look at humans. It's hard to believe that if we look at other species. Why is it true in humans? I guess it's because we have the power of language, and language creates culture, and culture makes the concept of the meme so powerful in humans, in a way that it is not clearly so obvious in other animal species.

*KH: Why did Dawkins invent that word 'meme'? Why didn't he just use 'culture'? Was it some sort of 'meme/gene' rhyming thing going on?*

**PC:** Yes, I guess it was. I think he wanted to use the scientific metaphor of the gene in a very vivid manner to show that some of the ideas behind the way in which the gene replicates could be seen in the way in which ideas replicate or culture replicates.

*KH: It's a complicated idea, and it sounds very 'spacey' – a very 'spacey' idea for a scientist to use – doesn't it?*

**PC:** It does, it does. And maybe Dawkins is straying way outside his area of expertise as an ethologist, as an animal behaviourist. I think

121

sometimes, in my conversations with you, I stray way outside my knowledge or expertise.

**KH:** *Excellent. It's what we aim to achieve.*

**PC:** Exactly. You know this whole business of evolution that I've been discussing with you is really a fascinating topic to me because I had a very interesting experience as a young scientist. I came across someone who was a so-called 'creation scientist'. This man was visiting from the United States and he gave a lecture at the university in which he showed that the earth couldn't be more than a few thousand years old. He presented a physics argument for this, the argument being that at the centre of the earth there is a lot of radioactivity which produces radon and helium that subsequently comes up from underground. And he argued that because the velocity of helium at the outer edges of the atmosphere wasn't sufficient to escape, if the earth was as old as the scientists were claiming, there'd be a lot more helium around. It sounded like a very good physics argument, and one of the professors in the university, a Christian, put to me the question, 'Well, what's wrong with this man's argument, Paul?' It cost me some time in the library to find the answer – it was before the days of the internet or Google and you had to work through the paper references. I found the answer, and the man was technically right, if he ignored the fact, apparently well known to atmospheric physicists, that every time there's sunspot activity, the exosphere of the earth heats up and all the helium goes into space. This is a very abstruse fact known by people who understand atmospheric physics.

So here was this creation scientist travelling the world, telling a story which he must have known was untrue. He must have known his argument had a fundamental flaw, but it served the propaganda of his argument. I wrote to him, of course, and presented the information that I'd found, and I've never had a reply. I guess I learned then the great difference between people who are *true* scientists – those who really seek to find the consistency, the truth and the rules that we explain nature by – and those who utilise parts of those rules for their own political or religious ends.

*KH: How do you know that the sunspots relieve us of the helium?*

PC: Well, of course we have marvellous tools now in terms of satellite tracking systems and detection systems, and we just know so much about the outer atmosphere as a result of the developments that have occurred since the 19th century.

*KH: Does the cause of creationism say 'Oh that's handy, isn't it, that's a handy little argument that's very hard to prove or disprove'?*

PC: Yes, but it's an argument that builds some consistency back into our understanding of the way nature is. I think that in science you have to have consistency between the rules, and when inconsistencies appear, then it's incumbent on science to seek out and find the problem and find out if the basic ideas, laws, rules, behind them are wrong.

# CHAPTER 11

# PSEUDOSCIENCE

Irrationally held truths may be more harmful than reasoned errors.
—T.H. HUXLEY

*KH:* *How are we to make sense of claims that eating genetically modified foods or living near power lines is unhealthy, or for that matter whether alternative medicine remedies can cure us? And how are kids expected to sort out evolution versus intelligent design, which is also claimed to be scientific?*

*PC:* I think that it is far more important to teach about what science is, and how it works, than to teach particular facts of science or outcomes from science. Of course, it's good to be taught those outcomes as part of education as well, but to get an understanding of what science is, and what it is not, I think is tremendously important.

*KH:* *Well, let's take the case of evolution, then, where creationists or intelligent designers complain that there are 'holes' in the theory of evolution. Why not teach that, then?*

*PC:* The fundamental difference in the way science operates and so-called 'creation science' or 'intelligent design', is that science, in the face of evidence to the contrary, moves on, acknowledging that the theory doesn't work. That's a fundamental difference between science and pseudoscience. Rules of evidence are absolutely crucial in science. You never hold on to some cherished belief in the face of evidence to the contrary. You say, 'No, I must have got it wrong, let's look at this again.' The difference in so-called alternative science or pseudoscience – and 'creation science' is an example, I guess – is that you hold on to that belief as a matter of faith, in the face of incontrovertible evidence to the contrary. For example, if creationists held to the position that the earth was 6000 years old, then they would be flying in the face of evidence.

127

*KH: But what about those who promote 'intelligent design'? Surely their perspective would be more acceptable to people like you?*

**PC:** But intelligent design requires a new law of nature. What is the intelligent designer and how was this designer created?

*KH: Well, why can't they put the question of who is the intelligent designer on the back burner, just as the aficionados of Darwin's theory of evolution put gaps in the fossil record on the back burner?*

**PC:** But the fossil links are being filled in all the time. And the important thing about science is that it also requires consistency with what we know. One of the features of so-called alternative science is that it does hold onto beliefs in the face of evidence and claims 'other ways of knowing', often invoking words like 'holistic knowledge'.

There are several recognisable characteristics of pseudoscience. First of all, appeals are always made directly to the media, rather than through the normal scientific discourse or through scientific peer review. Also, pseudoscience is generally based on some sort of anecdotal evidence or anecdotal experience. And then, there's always some conspiracy that's trying to suppress it.

*KH: Well, those nasty scientific peers!*

**PC:** For example. But to go on, any evidence for claims will be at the margins of detectability, the suggestions that power line emissions cause cancer being very much in that category. And some new law of nature will be needed to explain it, whether we are talking about colour therapy or cold fusion or homoeopathy.

*KH: Coming from a bloke who believes in quantum mechanics, that's a bit rich! What could be weirder than that?*

**PC:** No, no, the evidence for quantum mechanics is overwhelming! Plenty of experimental evidence for that.

I must tell you a little story. I grew up in Wanganui. Now, Wanganui was a hotbed of colour therapy. In fact, my father used colour therapy for his rheumatoid arthritis. I'm sure it didn't do him any harm. It didn't do him any good. I even had colour therapy used on me as a child. It didn't do me any harm or good, as far as I'm aware.

*KH: Nice-coloured shirt you have on.*

**PC:** Yes, well, it's a colourful shirt.

*KH: Well, these things do no harm.*

**PC:** That's very true. I think that a lot of things like, for example, therapeutic touch or homoeopathy are essentially harmless in themselves. There is no evidence they have any benefit whatsoever, and provided people don't use them as a substitute for seeking proper medical advice, then there's no danger in it.

*KH: Hang on a moment. If you accept the placebo effect, for example, which at the basic minimum you must, then why not?*

**PC:** Yes, absolutely. I quite agree with that.

*KH: So why not?*

**PC:** Why not indeed, if the claims are honest. I have no argument with faith healers, for example. Because faith healers openly offer the placebo effect. They say, 'If I take an interest in you, if I talk to you, I maybe pray or something, you will feel better or gain strength', and people do.

*KH: Maybe the placebo effect is the wrong expression, because the 'placebo' implies that it's nothing.*

**PC:** No, no. The placebo is something. Every doctor knows that. Every general practitioner knows that. And we know it ourselves. If we go to the doctor often our aches tend to disappear as a result of going to the surgery, even before we've got our prescription. We all know this is true and it is a well-measured effect. The disagreement I have with pseudoscience is not about trying to help the sick; it's about people claiming the language of science to justify what they do, but not accepting the operations and tests of science. And let's face it, homoeopathy is a multibillion-dollar worldwide industry.

*KH: Well if it works for the Queen it's all right for me!*

**PC:** There's nothing wrong with people spending their money on homoeopathy if they wish. But I do take issue with claims that assume the mantle of science without its obligations. As a physicist I've seen so many of these claims, whether they be antigravity machines, cars that run on water, perpetual motion machines ... The list is endless. None of these theories change when the evidence is against them.

*KH: So what's the evidence against homoeopathy? People say, 'Look, it works.'*

**PC:** That's the anecdotal point I made. You know, in medicine there have been many wonderful discoveries. There have been antibiotics, vaccination, chemotherapy and the aspirin. But the greatest discovery in medicine, greater than all those things, is the randomised, double-blind, placebo-controlled test. That is where you do a test and neither the people administering the test nor those receiving the test know if they are receiving the actual treatment or the placebo. It is double-blind and randomised. Only at the end of that exercise is the data collected and examined. You see if the treatment made a difference, and that's how medicine has to work. There is no scientific medicine that you buy from the pharmaceutical companies that hasn't been through such a test. It is required. But it is not required of homoeopathic medicine because people are literally buying either pure lactose, if it's in solid form, or pure water.

*KH: With a memory!*

**PC:** With a memory. You have to propose a memory, because the dilution is so great that not a single molecule of the proposed active ingredient remains. Now, there have been hundreds of tests of homoeopathy carried out worldwide. In fact, there is a million-dollar prize out there for anyone who can show that in a randomised, double-blind, placebo-controlled test that homoeopathy works, and every one of those tests has failed. No one has ever won the prize.

*KH: You mean absolutely no difference?*

**PC:** There is no difference from receiving a placebo. There is no difference that is apparent. Now, there's no harm in that.

*KH: No, there is no harm. Because people feel better.*

**PC:** But there would be harm if people were really ill and saw their homoeopath but didn't see their doctor. That would be harm.

*KH: So what about you? Would you use 'alternative medicine'?*

**PC:** I've had Reiki done to me by a friend.

*KH: What is that?*

**PC:** Well, the therapist's hands are run over your body without actually touching, and the 'aura of the body's energy' is manipulated. You close your eyes and these hands run over you. Actually, I found it quite thrilling.

*KH: Hmm...*

**PC:** There's a lovely story about that, about a little girl called Emily Rosa. Emily was a nine-year-old from Colorado who did a science fair project in 1996. She looked at these people who did therapeutic touch. She had a very simple little test. She made a screen so that the therapeutic touch person couldn't see her and they had to put their hands through two holes in the screen, and Emily would then toss a coin. Depending on whether it was heads or tails she would either put her hand under their right hand or their left hand, close but not touching, and they had to say which one it was. They all said before the test, 'Look, we'll know, we'll feel your aura, it will be so obvious.' Well, Emily did this test. She did 280 measurements on 21 separate touch therapists, and they got it right 47 per cent of the time. Fifty per cent would have been random. They did worse than that. Afterwards they said, 'That's not fair. We need to get to know you better. We need to get to feel your aura.' So they had some experimenting and felt her aura and then they did the test again and this time they got it 41 per cent right.

Emily was the youngest person ever to have an article published in a scientific journal. Her work was published in the *Journal of the American Medical Association*. It was an astounding project and it certainly raised some questions about therapeutic touch. It doesn't work and even a nine-year-old girl was able to show that. But people will still believe in it because it makes them feel good and that's absolutely fine.

*KH: I just wonder whether there's some way – and I'm trying to think of a way through here – whether the scientific method can ever apply to an individual because we are each so unique.*

PC: Well, we are. And I think there is something very interesting about human beings in that we have a thing called the belief engine. Some people call it the regressive fallacy. Humans naturally feel that if B follows A, then A must have caused B. If it doesn't rain for a long time and we do a dance and then it rains, well, the dance causes the rain. Humans have very special ways of looking at the world. We see patterns everywhere. We look at clouds and we see faces. The hunter goes into the forest. He sees a deer in the trees. He absolutely knows it's a deer. He fires. He kills his friend. But he really did see a deer.

There's a wonderful saying by the Chinese philosopher Lao Tse, in AD 600. He said, 'My axe went missing and my neighbour's son walked like a thief, spoke like a thief and looked like a thief. Then I found where I had mislaid my axe and my neighbour's son walked, talked and looked like an innocent child.' Isn't that so true?

*KH: Yes.*

**PC:** I mean, 'other ways of knowing' really gets us into trouble. Other ways of knowing is how people identified witches. So part of civilisation and part of the achievement of science is to break us free from these 'other ways of knowing', where our human belief engine lets us down so terribly badly and often with tragic consequences.

*KH: Here's the thing, though. Sometimes, by chance, science can be short-circuited. By accident. In other words, one minute science tells us that vitamin C staves off colds, then all of a sudden, no effect. One minute carrots stop you getting lung cancer, and next minute they increase the chance. Now, it wasn't homoeopathists who put out those stories! It was people slaving away in laboratories.*

**PC:** I can give you stories going in both directions. We once believed in phlogiston as part of burning.

*KH: Phlogiston? Sounds like a small state in the former Soviet Union.*

**PC:** Physicists also used to believe in the existence of the 'ether' in space. Phlogiston and the ether were scientific ideas that had their time and they were proven to be not consistent with evidence. They were ditched. Then there were other scientific ideas that started by being rubbished. Like continental drift, or black holes, or the Big Bang, and then ultimately the evidence showed that they were true, and we now believe them. So that's the point about the way science works – that it is able to ditch those things that are wrong and not cling on to them with religious faith. And, furthermore, those things that were previously rejected but which were found to actually fit the evidence, science is able to accept by saying, 'Oh, that was right all the time!' It's the flexibility in science, the ability to learn from the evidence, that makes its power, and that's what stands it apart from 'other ways of knowing'.

*KH: However, you are still talking about some ideal called science. And science is a house of many rooms filled by some completely cranky scientists who may not be completely ethical at all.*

**PC:** I totally agree. But science is not ultimately about the individuals; it's about the methodology. It's about the requirement of evidence and consistency, a process in which the chaff is separated from the wheat. Through the winnowing process, truth gradually emerges.

*KH: Eventually! After some of us may have died in the process.*

**PC:** Sometimes it takes a very long time. Look at the case of atoms, the idea that was rejected by many scientists for the whole of the 19th century. It wasn't until the early 20th century that the evidence was plainly there. People died in the meantime, but that's the whole power of science, that eventually truth will out. I think that is one of the most wonderful things about science.

Often people think of science as an 'establishment thing', or scientific medicine as an 'establishment thing', but the point is that, ultimately, these things are fundamentally revolutionary in that they allow for change. In science, for example, the youngest person can prove the eminent Nobel laureate wrong. In faith-driven thought that isn't the case, because it is a virtue to hold on to beliefs in the face of evidence, appealing to 'other ways of knowing'. That's the deeply conservative approach, in my view. Science is the flexible and revolutionary approach.

*KH: So you have faith in science, then?*

**PC:** I don't have faith in science at all. My experience of a lifetime tells me that the methodology of science has great power and great value. That is nothing to do with faith. Faith is completely separate from science.

*KH: Are you sure it wasn't the colour therapy that did it for you?*

**PC:** Maybe it did.

# CHAPTER 12

# RADIATION

No one that encounters prosperity does not also encounter danger.
– HERACLITUS

*KH: What is radiation?*

**PC:** It's really the transmission of energy by waves or particles. It includes things that we are all familiar with, like radio waves and ultraviolet light, which we worry about with sunburn. But it also includes things we are not so familiar with, like the radiation which comes from radioactive decay of atomic nuclei – gamma rays, beta particles, alpha particles – the sort of radiation which killed Mr Litvinenko in London recently.

*KH: From polonium-210! So what's the definition of radiation? For example, as opposed to sound waves?*

**PC:** Sound waves are a form of radiation. It's just that the energy is carried in a different way. Sound waves are caused by little pressure oscillations that propagate in the air, just as sea waves are little height oscillations that propagate on the ocean surface. Both carry energy. But, generally, when we are talking about the sort of radiation that we worry about, in terms of effect on human health, these are a special group. They are generally electromagnetic radiations, radiations that are carried by oscillating electric and magnetic fields in space; or they might be ionising radiations comprising particles, the sort that would be associated with radioactive decay, like alpha particles or beta particles, or they could be other charged particles that you get from cosmic rays hitting the earth or particle streams coming from the sun's surface. Some radiations we do have to worry about, especially those that can break chemical bonds. This can mean damaged DNA, induced cancers or other biological effects. But the whole field of radiation is so wide, between completely benign radiation that we needn't worry about at all, and radiation with which we must take enormous care.

*KH: What's completely benign?*

PC: Well, radio, of course!

*KH: Not what I've heard!*

PC: There are worries about cell phones, and there have been studies that suggest dangers, but most people might say 'Do we know of family or friends who have died from using a cell phone?' Excluding, of course, car accidents caused by using one while driving!

*KH: So it ranges from harmless to harmful. What makes it harmful? That it's strong enough to break the chemical bonds in our body?*

PC: Yes, it's a very interesting point that really comes down to quantum mechanics. Take electromagnetic radiation, which ranges from very long wavelengths, like radio waves, to very short, like gamma rays. What Einstein pointed out was that the wavelength of the radiation determines how much energy can be transferred when that radiation interacts with matter. He called that lump of energy the photon, and that photon energy is very critical when we think about human life, because if the photon energy is enough to break a chemical bond, then the radiation can damage the molecular structure in humans. The potential damage point comes just at the violet end of the visible spectrum. Visible light is pretty harmless. It can't do much in the way of breaking bonds (except those so weak they are likely to break of their own accord). But once you get beyond violet into the ultraviolet, then you start to have enough energy in the electromagnetic wave photons to break the bonds.

*KH: That's life, isn't it? The stuff you can't see does the most damage!*

PC: Fortunately, on earth, we are mostly protected from the ultraviolet by the ozone layer at the top of the atmosphere. It cuts out most of the ultraviolet, though a little bit gets through – enough to give us sunburn – and in some places on earth, like Australia and New Zealand, quite a bit more gets through. But we humans can live with a bit of ultraviolet and we've evolved to be that way. And, interestingly, there are some beneficial aspects to ultraviolet. It produces vitamin D. It's actually quite good for the skin in small amounts. If you don't get enough of it, if you never go outdoors, you'll never be healthy. So humans have evolved to live with the dangers and the benefits. But once you get to shorter and shorter wavelengths, the far ultraviolet, to X-rays and then into gamma rays, then you are really looking at potential danger.

KH: *So gamma rays are very, very high energy and very, very short wavelength?*

PC: Yes, and only generated out of nuclear decay, except in some rather exotic process in deep space. And gamma rays reach the surface of the earth because of cosmic radiation. We've evolved to live with it.

KH: *And when you say short wavelength, you mean quite literally when you draw the waves, the distance between the crests is small. Is that a metaphor or is that true?*

PC: No, it's true. The wavelength is a measurable thing.

KH: *So, coming back to radio waves then, I saw a story the other day about mobile phones and it quoted an emeritus professor of physics in Britain, Lawrie Challis, who says that mobile phones could be the cigarettes of the 21st century. He says that they've been relatively unproven in terms of brain damage or tumours so far, with a very slight association between the risk of brain tumours and the use of mobiles for more than 10 years. But he says that things take more than 10 years to show up, so we can't tell.*

PC: There was also a report recently in *The New Zealand Herald*, related to another study. I think the study that Lawrie Challis referred to was Swedish and this latest one, reported by the newspaper, was a Finnish study, again relating to possible effects for people who have used a cell phone for more than 10 years. So it's very interesting and I'm sure quite a lot of people will be alarmed by that.

*KH: The first paragraph of that* Herald *story says that people who use mobile phones for more than a decade are far more likely to grow brain tumours on that side of their heads.*

**PC:** Well, I was fascinated by that story, especially when I saw this expression 'far more likely', so I decided to find out what the study was. I went to the *International Journal of Cancer* on-line and found the article. I'd like to read you the abstract. It says: 'For more than 10 years of mobile phone use reported on the side of the head where the tumour was located, an increased odds ratio of borderline statistical significance was found.' Now that means there was a borderline indication that you might have more chance if you used a cell phone for 10 years, than if you didn't. It goes on to say, 'Although our results overall do not indicate an increased risk of glioma in relation to mobile phone use, the possible risk in the most heavily exposed part of the brain with long-term use needs to be explored further before firm conclusion can be drawn.' Now, I have to say that's a big difference from 'far more likely', as in the newspaper article.

*KH: Well, it is a big difference. But also, the second paragraph of the newspaper article says that results from this European study of more than 5000 people showed that mobile phone users were 40 per cent more likely to develop a type of nervous system tumour near their phone ear. Now how does that correlate with what the report actually says?*

**PC:** It's very interesting, isn't it? I went and looked at the same studies that Challis was referring to, the Swedish study and some other Danish studies, and let me give you the information I found. One enormous study involved 420,000 people. There were 14,000 cancers within that group of 420,000 people, and the conclusions of that Danish study were 'we found no evidence for an association between tumour risk and cellular telephone use, among either short-term or long-term users'; and so on, also stating 'any large association of risk of cancer and cellular telephone use can be excluded'.

Then there's another study, the Swedish one that Challis refers to, which says 'our findings do not indicate an increased risk of acoustic neuroma related to short-term phone use after a short latency period. However, our data suggests an increased risk associated with mobile phone use of at least 10 years duration.' Now here's the interesting thing. That Swedish study was focusing not on cancers or tumours *generally*, but on one very rare type of tumour called an acoustic neuroma. Another study published in the same year in the *American Journal of Epidemiology* states 'there is no association between long-term cell phone use and acoustic neuroma'. Same-sized sample, but done in Denmark rather than Sweden. So what are the public supposed to make of this?

*KH: Yes! What on earth does it all mean?*

**PC:** Well, let me attempt to explain. All this is based on something called epidemiology. And epidemiology is not about finding mechanisms and cause and effect, but about looking at correlations in a large sample of people. To put this in simple and familiar terms, there is a type of epidemiology that everyone understands. It's called opinion polling, political opinion polling. You say, 'Let's get 1000 people and see how they vote and see what it correlates with.' Imagine this hypothetical opinion poll. Suppose a sample of 1000 people who didn't use cell phones suggested equal numbers would vote National and Labour. Whereas in the other sample, where people used cell phones, more people supported National than supported Labour.

*KH: Cell phones encourage National Party support!*

**PC:** Now, no rational person would draw that conclusion. There's an obvious confounding factor. It may be that people who use cell phones are more prosperous and maybe they tend to vote for the right rather than the left. One of the problems with epidemiology is the chance of a confounding factor. If there's a correlation between two things, it doesn't mean there is a cause and effect. The second problem with epidemiology happens when you go to very small samples. Suppose amongst this 1000 people, it is found that there are two voters for the

McGillicuddy Serious Party but amongst the cell phone users there are five voters for McGillicuddy Serious.

*KH: A 150 per cent increased chance!*

**PC:** That's a huge effect. Obviously, for McGillicuddy Serious, there must be a huge effect from cell phones. But we know that's silly. The problem is that when we come to the small numbers of votes for the minor parties, the errors become very big. So here's the risk. If you do a study with cell phones and you have 420,000 people, and then narrow it down to 14,000 who have tumours, and then within that the 100 who have acoustic neuroma, and you narrow it down to only the people who have used cell phones for 10 years, you are down to about a dozen people. With such a small sample, given the natural variations which will occur, the chances are you are going to see a slight effect. And that's the risk.

*KH: So that's how we should read that paragraph that I read out earlier, 'Long-term mobile users are 40 per cent more likely to develop the type of tumour . . .'*

**PC:** That's indeed what they found. But then what happens is that you have to give what is called a 95 per cent confidence level. So you've 40 per cent more chance but the confidence level takes you somewhere between a 10 per cent and a 70 per cent effect, it's somewhere in that range. This is, as we say in physics, really 'down in the noise'. This is something that we really can't be confident about.

*KH: But the fact is, as Professor Lawrie Challis pointed out, the exposure we get from mobile phones is tens of thousands of times more than we get from a television set or a mobile phone mast. So I'm thinking, if it's that much more, surely there has to be some side-effect, if radiation is bad for us.*

**PC:** Well, we do have to be very careful about the type of radiation we are talking about. We generally divide radiation into two classes, what we call ionising radiation, and non-ionising radiation. The difference is that ionising radiation has enough energy to break chemical bonds, to change molecular structures. It can break DNA, for example, and cause a cell to malfunction, or cause a genetic defect. Non-ionising radiation can't do that. No matter how intense it is, no matter how much radiation there is, it cannot break the bond, it cannot cause a change in the chemistry. What it can do, however, is generate heat. That's how a microwave oven works. Microwaves are an example of non-ionising radiation. They make things hot. With enough non-ionising radiation, you can cook. So the thing about radio waves, the radiation that cell phones generate, is that they are a form of non-ionising radiation; so at any dose, they can't break chemical bonds, but they can cause a warming effect.

*KH: A simple warming effect? To the extent that 'if men spoke to their testicles with cell phones, they would become impotent', that sort of warming effect? I should clarify that, but I'm not even going there!*

PC: Well, our body is very well attuned to warming effects. If my hand is too close to the barbecue, I pull it away. So, as far as we can understand, the only effect of radio waves on the body would be a small warming effect. And in the case of cell phones, it is so minimal we would be hard put to measure it. But that doesn't mean we should rule out other possibilities.

*KH: It also doesn't mean that it's harmless, because, if a repeated warming effect is directed at a particular part of your brain, then . . . what's the result?*

PC: But in the case of cell phones, with power levels of less than half a watt, this warming effect is almost vanishingly small. And, we have had vast experience with humans being exposed to radio waves for well over a century.

*KH: I know, but then, that's what people say about chemicals, that this bit doesn't hurt and this bit doesn't hurt, but people who are opposed say, 'Yes, but add them all up.'*

PC: It pays to be precautionary. And one should take this very seriously, which is why there is a lot of research going on in this area. But I have to say that there seems to be no convincing evidence, so far, that cell phones cause a health problem, even after 10 years of use. At least, that's the way I would read the epidemiology. It would be a big stretch to say, as Lawrie Challis did, that mobile phones could be the next tobacco. That statement would surely have alarmed many cell phone users. On the other hand, I might enjoy a little mischievous pleasure at the alarm of some cell phone users. Especially those who get on the train or the bus and have conversations in a loud voice.

*KH: Well, unless it's me doing it. And, of course, I'm having a very important conversation.*

**PC:** They are antisocial things.

*KH: So, if there is such a big difference between ionising radiation and non-ionising radiation, why are both called radiation?*

**PC:** Because they are. The word radiation is used generally to describe the emanation of energy from some source. There is a very natural fear associated with the word because it's often an energy we can't see or feel or sense in any way directly, except for that narrow portion of the electromagnetic spectrum we call 'visible', where our eyes are attuned to see light. And, because some radiations, in particular ionising radiations, can be very dangerous, there is a natural anxiety on the part of people concerning all radiation.

*KH: Yes, and people worry, for example, about things like power lines.*

**PC:** Power lines do produce electromagnetic waves, of ultra-low frequency and extraordinarily long wavelength, thousands of kilometres in fact. And the question is, do these waves that come out of power lines affect people?

*KH: What about the pylons? With a confluence of power lines, would a whole bunch of long-wavelength waves add up to a short-wavelength danger? Do you understand that question?*

**PC:** I do understand and it's an excellent question. That is, if you have enough radiation of very long wavelength with very low energy photons, could they do damage?

**KH:** *Yes!*

**PC:** This is the astonishing thing that we know from quantum mechanics, that the amount of energy transferred when electromagnetic radiation interacts with matter is not associated with how much radiation there is, but with the energy of the photon; in other words, how much energy can be transferred in these lumpy events as the photon is absorbed. The intensity of the radiation merely tells us how many photons there are, not how energetic each one is. And the photon energy is all to do with wavelength. If the electromagnetic radiation is not in the ultraviolet, or shorter wavelength, then it won't break a bond. As with the radio waves, such low-frequency electromagnetic waves deposit their energy as a mild warming effect, though in the case of the radiation from power lines, it's so small it's immeasurable.

**KH:** *So does that mean that these radiations must be safe?*

**PC:** Nothing can be proven safe, only unsafe. The fact that such electromagnetic radiation won't break a bond is not to say that the electric and magnetic fields associated with the waves couldn't cause some other biological effect.

**KH:** *What do you mean, 'other biological effect'?*

**PC:** Well, for example, our cells have ions moving across their membranes, they conduct electricity locally. So an electric field might affect the conduction process and the way cells function. So it pays to be precautionary, it pays to look very carefully at any possible effects. And the evidence so far seems to suggest that there is no danger. And this is a matter that has received enormous attention, and which has been subject to intense scrutiny by scientists and health researchers, with enormous epidemiological studies, in the UK and Europe, and extensively in the United States. These show there is not a significant measurable effect.

**KH:** *Childhood leukaemia is the* big *concern, isn't it? And people seem to be indicating that there are clusters; for example, the recent British study.*

**PC:** Yes, this was a report coming out of Oxford University and published in 2005. They looked at 10,000 cases of childhood leukaemia in the UK and came up with a strange result. They correlated leukaemia cases with where children were born. Not with where they grew up or lived, but where they were actually born. And the study found a slight preponderance of childhood leukaemia for children who were born within 600 metres of high-voltage transmission lines. But the preponderance was marginal. We are speaking here about a few tens out of 10,000 cases, and where the levels of confidence aren't all that great. What they came up with was 50 per cent greater chance of childhood leukaemia, but within the range of a 10 per cent to 100 per cent extra chance.

**KH:** *You mean, you don't know whether there is an effect because the statistics are so small?*

**PC:** It's a weak effect, a bit like that example of the opinion poll accuracy as regards the very minor parties. To put the figure in context, it represented a possible five extra cases of childhood leukaemia a year in the whole of the UK. If that were real, then those lives matter and we would take it very seriously, but the problem is, the indication of any effect is just too unreliable. And then there are those confounding factors of epidemiological studies. A correlation doesn't mean a cause and effect, and when the indications are this weak, that's a double problem.

I should point out that in the extensive studies undertaken in the US over a long period of time, involving the National Institutes of Health and the US National Academy of Sciences, the final conclusion was that no observable effect was found from power lines. My own reading is that the scientific evidence suggests we don't have much to worry about from power lines, at least from a health effect resulting from radiation.

*KH: Well, let's talk about what we do have to worry about, like polonium-210.*

**PC:** Oh, some radiations are very dangerous, nuclear radiations for example.

*KH: And X-rays?*

**PC:** X-rays can also be dangerous.

*KH: Do you know what? When I was a child we had this shoe shop and for a treat on a Saturday morning we used to go in and put our feet inside this little machine and you could look through the window and see X-ray images of the bones in your feet.*

**PC:** You don't see those around any more. They're banned! That was an irresponsible use of X-rays.

*KH: But people presumably then said, 'Oh it couldn't possibly cause any harm because it's feet.'*

**PC:** Yes, but the evidence became clear that it was dangerous. We've learned a lot about ionising radiation since then.

*KH: Have I been damaged by that?*

**PC:** I don't think you probably have, but in a large population of people you might find that some could suffer ill effects. For any one individual, the chances of ill effects would probably be small. But it is an irresponsible and unnecessary use of ionising radiation, to use X-rays that way.

But there is a context in which we must view all exposure to ionising radiation. We as humans are exposed to ionising radiation all the time, because of cosmic rays coming from outer space, because of radon coming out of the ground, and even because of some naturally occurring radioactive elements that are inside our own body. To set that context more clearly, when you have a chest X-ray, you get the equivalent of your normal natural annual dose of ionising radiation. That annual

147

dose, something around 2 milli-Sieverts, is a sort of reference level. We don't want to exceed that too much. The annual natural dose is not terribly dangerous, and one can estimate the risk of cancer of genetic abnormality for every extra milli-Sievert you have. It's about 0.004 per cent, a very small effect. But still, it's what we've evolved to withstand, and we don't want to unnecessarily exceed that. But if you are working in an industry where there is a large amount of radiation, you face an occupational hazard. If you are an X-ray radiographer, for example, you don't want to have a major increase over the natural background.

*KH: It always strikes me as kind of sinister, though of course I understand the reasons for this, that while one is lying there being X-rayed, everyone else leaps out of the room.*

**PC:** But radiographers do it every day. You get it once every couple of years!

*KH: I know, I know.*

**PC:** Another occupational group exposed to radiation is airline pilots. If you travel at high altitudes, at 10,000 metres, on that trip from Auckland to London you are exposed to additional cosmic rays. More than if you remained at sea level. One such trip is just an extra 10 per cent of your annual dose. But if you are a pilot doing it many times a year, you could easily get two or three times your annual average natural dose.

*KH: And what are the consequences?*

**PC:** Probably a quite small increased risk of cancer. Very small, but finite. Yet you could easily have this higher dose living in some parts of the world where natural radiation doses are much higher than others. Pilots know about that. It's part of the risks of their profession.

*KH: Maybe that's why they retire young?*

**PC:** More likely, it's problems with eyesight!

*KH: What about astronauts?*

**PC:** Astronauts can be exposed to much greater danger, being very exposed to cosmic rays, and, especially if there happens to be a solar flare, to ionising particles streaming out from the sun. They do have to have protective clothing and they do face risk.

*KH: And so we come to the really dangerous, the polonium-210. Bad stuff to eat!*

**PC:** Do not eat polonium-210! The Litvinenko poisoning was extraordinary.

*KH: It was extraordinary. Did you see the Panorama documentary about it?*

**PC:** No, I didn't.

*KH: They showed how, if it had been administered in a glass of water, or cup of tea, as the supposition goes, it would be possible to find traces of it everywhere the liquid, or the person who had drunk it, had touched.*

**PC:** It has a half-life of about 130 days. It lingers around for quite a while. Polonium-210 is an alpha-ray emitter, alpha rays being the charged nuclei of helium. Practically no gamma rays are emitted from polonium-210, just these energetic alpha rays, which stop in a very short distance, much less than the thickness of a human hair. Hence, they are very difficult to detect from outside the body, once the polonium is ingested. The short stopping distance means that the alpha particles do damage right at the site of the cells in the vicinity of the polonium atom, but with no long-distance radiation coming out, as one would have if the polonium emitted gamma rays. If you ingest this stuff, it's stuck inside you tearing into your cells, your DNA, your proteins, whatever is nearby.

By the way, on terminology, the polonium is the radioactive material, the alpha particles are the radiation. One can be 'contaminated by radioactivity', but one is 'irradiated by radiation'. Often we hear the expression in movies or read in newspapers that someone was 'contaminated by radiation'. That makes no sense. It's the radioactive material which lingers to do its continuing damage, not the radiation, which gets one shot to do damage and then is spent. This is much more than playing with words. Understanding that terminology can be a matter of life or death!

But to return to polonium-210, let me tell you how dangerous that is. It is, I think, the most poisonous material known. If you drew a full stop with a ballpoint pen as small as you could make it, so that you could just see the dot, then if that dot were polonium-210 it would be enough to kill 200 people. It is five million times more toxic than hydrogen cyanide. Fortunately, we don't have to worry too much about polonium-210. It's a rare isotope which can only be produced in a nuclear reactor, and doesn't occur naturally on earth.

*KH: So, if Litvinenko had drunk it in a cup of tea, for example, and I had kissed him after that, would I die?*

**PC:** Not a good thing to do, Kim. Very dangerous. Yes, a fair chance of a fatal dose being picked up that way, especially since a full-stop-worth is enough for 200 people!

*KH: I don't want to be too gruesome about it, but an autopsy would have shown what?*

**PC:** Severe cell damage. Severe DNA damage. And since the polonium chemically binds to certain places selectively, certain organs selectively, anyone familiar with this toxicology would know which organs to assay. The tell-tale fingerprint of polonium-210, or indeed any radio-active material, can always be found by examining the precise energy spectrum of the radiations coming out. But as I said, what makes this material so difficult to detect initially is that there are no long-range radiations escaping from the body. Only a biopsy or autopsy could get at the material needed for the assay. The short-range stopping of alpha rays makes polonium-210 both lethal and hard to detect.

*KH:* *A very weird way of killing someone.*

**PC:** Very weird, very cruel. We are all educated about polonium-210 now, something we knew little about before this.

*KH:* *Something else we have to worry about!*

CHAPTER 13

# NANOTECHNOLOGY

Things on a small scale behave nothing like things on a large scale.
– RICHARD FEYNMAN

*KH:* Nano?

**PC:** Nano means small. It comes from a Greek word meaning 'dwarf' but in scientific terms it means one-billionth. So when we say 'a nanometre', we mean a billionth of a metre. That length is hard to get your head around, but I can give a few examples. A human hair, which is about 0.1 millimetres thick, is 100,000 nanometres. A red blood cell, something pretty small but nonetheless something we have some idea about, is around 7000 nanometres in diameter. An atom, one of the smallest things we can think of, is just a bit smaller than a nanometre. So molecule size is measured on the nanometre scale. A lot of the organelles in living cells are on the scale of tens or hundreds of nanometres. So it's a very useful length scale when we are talking about atoms or molecules or very small biological objects.

*KH: Nanotechnology, the concept of it, has been around for a while. Richard Feynman made his seminal speech in 1959, and everybody cites him, whenever they talk about nanotechnology, to gauge how far we've come away from his vision. What did he imagine?*

**PC:** It was really quite a fantastic idea when he first proposed it. He thought of the possibility of being able to manipulate individual atoms.

*KH: Why not?*

**PC:** Why not! We didn't have the technology then. But if we could manipulate atoms one by one, we could for example have, on the head of a pin, an array of a million by a million atoms, enough to store an entire *Encyclopaedia Britannica* worth of information. On the head of a pin! If we could write with individual atoms.

**KH:** *Can we do that?*

**PC:** In principle, at least since the 1980s, when we had this new device invented, the scanning tunnelling microscope. This enables us to 'see' atoms, and to pick them up and move them around.

**KH:** *With what?*

**PC:** With a very fine metal tip which could be manipulated by very delicate 'piezoelectric transducers'.

**KH:** *So essentially this is the nano-robot.*

**PC:** This is the nano-robot in a sense but, really, the device only allows you to do this with a few atoms at a particular place. You couldn't mass-produce with a tool like that. You could go to a particular place on the surface of some material. You could pick atoms up and move them around, but it's not a manufacturing technique.

**KH:** *So what is the utility of that?*

**PC:** The utility, first and foremost, is being able to see individual atoms, to start to see behaviours that weren't previously visible. But, more importantly, it gave us a way to think about how we could manipulate structures, on a scale that we couldn't previously have imagined. And that brought into being the idea of nanotechnology as a form of engineering, where we could create structures at the atomic or molecular scale.

**KH:** *But we couldn't create structures that wouldn't naturally create themselves, could we? I mean, the atoms have to have some affinity.*

**PC:** Indeed, but we can use the atoms' affinity, their 'stickiness', to make new engineering structures. One of the first spectacular examples was the team led by Don Eigler at IBM labs, which used individual atoms to write the letters I B M on a surface. It was a bit of a gimmick but it helped create an idea in people's minds about what might be possible.

*KH: So what are we able to do at this point?*

**PC:** At this point, we are not able to do a great deal more, atom by atom. But at a slightly larger nano-length scale quite a lot is being done. What is driving the technology is the electronics-device industry. We are so used to the idea that every year computers get more powerful, that we have more memory on the chips, that the computer processors go faster and faster, and that's all being driven by making structures smaller and smaller. We are now passing the limits of what can be achieved by optical methods, with lithography, to create chip features. The finest line we can currently draw on a chip is about 100 nanometres in thickness. To go lower than that, to put even more 'memory on the chip', to make the computers go even faster, we've got to be able to do nanotechnology and nanoengineering.

*KH: And the only way to get smarter is to get smaller?*

**PC:** Exactly. And if computers are going to continue to get faster as we head into the future, we have to be able to improve those technologies. The industry has laid down an 'IT Roadmap', so that by 2015 we need to be working at the 20-nanometre scale feature size, instead of the 100-nanometre scale. This requires new lithographic technologies, a new paradigm in terms of the way in which we make chips, 'top down'. That's one factor driving all this. But there is also another approach, the so-called 'bottom up' method, the way nature does it. Nature makes nanostructures building up from the lower level, and there is an enormous effort going into that, to see how we can create new materials and structures by assembling atoms and molecules from the bottom up, rather than engineering from the top down, as we do with lithography.

*KH: And creating materials that have hitherto not been known.*

**PC:** Yes, maybe new molecular transistors, or new molecular motors. But as a simple example of a new nanomaterial we are all familiar with, we have the nanoparticles that are the active ingredient in marvellous new sunscreens with very high protection. These particles, of zinc oxide or titanium, in the cream are totally transparent to ordinary visible light, because of their size, but they are highly reflecting to ultraviolet. That's a nanotechnology that seems very mundane, creating a useful product, and one whose nanoscience basis we don't even think about.

*KH: At the computer end, getting smarter and smaller, that's commercially driven.*

**PC:** Yes, indeed.

*KH: I mean, there are people who would say, 'Why?'*

**PC:** Why go faster?

*KH: Why do we need to go faster?*

**PC:** Well, we don't, in a sense. The faster computers go, and the more powerful they are, the more we have to run around in circles doing more and more things in our lives.

*KH: Yes. And I think that scares people.*

**PC:** I quite agree. And I think that there is a general feeling that science is somehow driving changes in all our lives that we don't have control over. This has happened before. Look at the way the contraceptive pill has had such a huge social impact. We've got used to that. But every new development produces some impact.

*KH: Unintended consequences.*

**PC:** Unintended consequences, and ones which scientists themselves can't foresee, that ultimately the public has to make decisions about, concerning how we use that science.

*KH: One of the issues about nanotechnology is the need for replication, at some level. Self-replication – can you explain that?*

**PC:** Self-replication is a worry that some people have had about these technologies. It really comes from the fact that the ultimate nano-technology is nature itself. If you think about a living cell, it contains within it all these organelles, marvellous little nanomachines all doing their particular jobs. And, of course, a cell self-replicates, and I guess there is a fear that the people who are driving these changes in nanotechnology might want to make little machines that mimic this, that can also replicate. This has led to a concern about something called 'grey goo'.

*KH: A scientific expression?*

**PC:** Not really. It was coined by Eric Drexler and picked up on by none other than Prince Charles, who expressed concerns about being swamped in grey goo.

*KH: Drexler was in a big debate with Richard Smalley, a Nobel prize winner. Smalley was saying 'Pull up, pull up, this is nonsense'.*

**PC:** Yes, and in fact Drexler has pulled right back from those concerns. It was really a 'science fiction' worry, that Drexler has now acknowledged is not important, that self-replication is not an issue. I don't think that any serious scientist would say that self-replication of nanomachines is even on the horizon, that it is not something that we can even envisage.

*KH: Well, why not? Because if you are going down to the atomic level, you can go down to the cellular level, and cells self-replicate. Why can't you engineer something that does that?*

**PC:** There are two principal reasons. The way cells replicate is completely 'bottom up'. They assemble atoms and molecules together in a way that is encoded by the DNA. The way that we envisage nanomachines being made is essentially top down. Ultimately, when we move the atoms around, we get down to a length scale at which it becomes impossible to do things with sufficient accuracy. This is because at the very small, the atoms are jiggling around with their thermal motion. That's one technical reason. And the second technical reason is because of 'stickiness', what's been called 'thick sticky fingers'.

*KH: Fat sticky fingers. That's right.*

**PC:** So the device that moves atoms around has to be on the nanoscale size, and it becomes difficult to disengage the structure you are making from the machine that is building it. So I don't think a machine can make itself at the nanoscale in that way. Nature works in an entirely different way, in a way that is far more complex, and in a way that we don't really understand. We don't yet really understand how a living cell operates, how the replication processes work. We understand in a crude way but not at all in the detailed way needed in order to try to imitate in synthetic engineering.

*KH: And so you're suggesting that nanotechnology has uses in computer science, as opposed to health.*

**PC:** No, I think there are tremendous implications for health, and that's where some of the ethical issues will arise. An example would be our being able to interface electronic systems with living cells. Nanotechnology will enable us to cross the bridge between silicon and wet tissue, meaning that if you wanted to make an artificial eye for someone who was blind, for example, then the eye would be some synthetic or engineered device a bit like a modern digital camera, but it would have to interface to the brain, so that we would have to connect the electronics and the nerves. Crossing that bridge between electronics and living tissue is something that nanotechnology can do, in principle, because it works at that level of size necessary for that interface. I think that one of the things we are going to see is a remarkable development in prostheses. We will see wonderful hearing devices, but the challenge, and not far away, is that we might succeed in making artificial sight. Then the next step is that we might be able to connect the brain itself to an external computer. It's not such a fantastic idea, an interface that goes directly from our brain to the most powerful computer, enhancing our own thinking power. That raises enormous ethical issues.

**KH:** *Like?*

**PC:** Like who gets access to this.

**KH:** *Whose brain are you going to enable?*

**PC:** Exactly. And if we can restore sight to the blind, then why not, for example, allow people to extend their already perfect sight to the ultraviolet, or the infrared? Creating a new 'super capacity' in humans. A good example of that ethical issue might be seen in sport. We have a concern about performance-enhancing drugs, and we've made a decision as an international community that we won't allow that in sport, that we don't want that sort of increased performance. I think we may have to make the same sorts of decisions about how we want to use new technologies that enhance human capabilities in other ways.

**KH:** *It's always that grey area that worries us. If somebody has an artificial hip and still manages to be the long-jump champion of China, that's fine, but...*

**PC:** There are no simple answers. And in the end, the public themselves have to make the decisions.

**KH:** *It all comes down to that question of what it means to be human.*

**PC:** Absolutely. Extraordinary ethical issues are going to arise as a result of new science, and we will see this in other areas, in terms of drugs that enhance the brain in certain ways, that enhance memory. How do we use the advances, when these things can change our very human nature?

**KH:** *One of the people who came out against nanotechnology was Bill Joy, who wrote about it in* Wired *magazine. His concern was what, in the main?*

*That it was going to lead to an exponential avalanche of stuff that we couldn't control?*

**PC:** That's right. We often talk about nanotechnology as being a 'convergence' science in that we see little incremental changes, but suddenly something switches on in terms of outcomes. Nanotechnology is working at the interface between physics, chemistry, biology and engineering, and that's an interface where there are so many potential outcomes that we can't envisage. The other word that is used is 'upstream'. It's a new science and there will be big impacts. But it's very hard to look ahead and see what they might be. Looking back at our own lifetimes we can see how science has brought about changes that have been dramatic and unforeseen. We know that this new science will have impacts that are unforeseen. But how can we stop it? How can we stop, and why would we stop, any advance that has the potential also to improve people's lives enormously?

**KH:** *Is it that you don't think that amount of foresight is possible, or that you think that that amount of foresight is not a good idea?*

**PC:** No. I think that foresight is a very good thing, but foresight is difficult. My view as a scientist is that, ultimately, these questions have to be decided by the public at large. Of course, the scientist is also a citizen and so can't avoid also facing up to the ethical issues. But the scientist's role is also to try to look ahead, to flag concerns, and to engage as openly as possible with the public in talking about science that is on the horizon, and things that may be happening. We've seen some disasters in terms of the miscommunication between scientists and the public, in areas such as nuclear power where the industry became inextricably associated with military uses, poisoning people's attitudes towards nuclear power.

**KH:** *Well, there's also the problem of getting rid of the nuclear waste.*

**PC:** Difficult, but there are new technologies, such as encapsulated fuel, so that the waste is much easier to handle, but people just don't want to know and I can understand that because of the history of that industry. Genetic modification is another, where some new technologies have been associated with companies that haven't shown much openness or concern for public perceptions. The key to dealing with all these issues is the maximum degree of openness on the part of scientists and industries that are working with these new technologies.

*KH: There are two extremes of view, though, aren't there? One's the apocalyptic view that hell in a handbasket is where science is taking us and the other is the Ray Kurzweil view of nanotechnology; for example, he says that by 2020 molecular assembly will provide tools to combat poverty, clean up the environment, overcome disease, extend human longevity and many other worthwhile pursuits. Well, clean up the environment. Let's go for it! You know, there's a kind of exaggerated view of what science can do. Don't you agree?*

**PC:** There is very much an exaggerated view, and you can exaggerate the dangers as much as you can exaggerate the potential benefits. But we just have to look at our own individual lifetimes to see how much science has changed the way we live. We live longer, we live healthier. We have less drudgery in our lives. At the same time these developments have brought more pressures in our lives. We are more connected in a way that is good in terms of information sharing but which makes us all run around a lot faster and worry about things that are happening on other parts of the planet. So it's clear that these advances bring with them problems as well as marvellous benefits. We wouldn't want to turn our back on those benefits. We really wouldn't want to go back to a world as it was at the beginning of the 20th century. Certainly I wouldn't want to go back to it. And it's not a question of us even being able to control it. Science is going to advance because it is part of human nature to want to push back the boundaries of our knowledge, to push out and learn more and more. We shouldn't fear that. But we all need to try to be aware of what is happening and to express a view about how science is to be used.

# CHAPTER 14

# THE CLIMATE OF PLANET EARTH

*My candle burns at both ends;*
*It will not last the night;*
*But ah, my foes, and oh my friends –*
*It gives a lovely light.*
—EDNA ST VINCENT MILLAY

**KH:** *I want to quote something, Paul.*

I expected science to banish the evils of human thought, prejudice and superstition, irrational beliefs and false fears; I expected science to be, in Carl Sagan's memorable phrase, 'a candle in a demon-haunted world', and I am not so pleased with the impact of science. Rather than serving as a cleansing force, science has in some instances been seduced by the more ancient lures of politics and publicity.

*The quote comes from Michael Crichton's book* State of Fear, *a blend of fact and fiction which attacks what he sees as the flawed science at the base of the global-warming fears. So what he's saying is that science is being exploited by scare-mongerers. What say you?*

**PC:** Well, he has some interesting points to make in that book. The book is a good read; quite long, but you can rattle through it in a few hours. He paints environmentalists as psychopathic killers, which I found curious. But Michael Crichton does have a point in that he worries that different standards of truth or verification are applied in different areas of science. He is a medical doctor, and in medical science we apply enormously rigorous standards of testing new drugs by double-blind methods. But a lot of areas of science don't have the same rigour. Sometimes experiments are done and never tested again.

**KH:** *So does he have a point about climate-change scaremongering?*

**PC:** Perhaps. But I have to say I don't really agree with Michael Crichton on this. I especially don't agree with him, having read much of the report by the Inter-governmental Panel on Climate Change –

*KH: Henceforth to be called the IPCC...*

**PC:** Yes. IPCC came out of a United Nations coordinating exercise and it involved two thousand scientists from a hundred countries, although not all of them agree with all of the report. Still, the report was a thorough study: in-depth and peer-reviewed. Of course, Michael Crichton says that science is not about democracy. It's not about consensus. It's not about how many people say that something is right. I agree with that. It could be that a lot of people say something and they're all wrong. But unfortunately there's nothing in Michael Crichton's book that really provides any better scientific insight than IPCC. In fact, if you look at the scientific arguments he produces to try and debunk the matter of global warming and global climate change, they're very selective and not too convincing.

*KH: Let's get back to some of his criticisms after we've been through the basics: what is the greenhouse effect?*

**PC:** The greenhouse effect is something that actually keeps planet earth about 33 degrees warmer than it would be if we didn't have the protective atmosphere around us.

*KH: So there's nobody who disputes the greenhouse effect?*

**PC:** No, and the greenhouse effect is a wonderful thing. Without it, the average temperature on earth would be $-18°C$...

*KH: We like the greenhouse effect. So all scientists believe in the greenhouse effect?*

**PC:** Well, we know it's there.

*KH: So how does it work?*

**PC:** We know that warm objects emit radiation. If I get a piece of iron and heat it up it starts to glow – it glows initially reddish and if I get it hotter it glows orange, and as I get it hotter and hotter it goes yellow, and eventually it gets white hot. So the temperature of the object determines the radiation colour. Now, the surface of the sun is very, very hot, 6000 Kelvin, and that's why it's white: it's emitting energy at about half a micron wavelength. That wavelength is exactly where, by no accident, evolution has caused our eyes to be the most sensitive. That's what we call the 'visible' part of the spectrum of light.

By contrast with the sun, the earth is very much cooler and it emits its own radiation at a much longer wavelength. It's much, much further down in the spectrum, with a wavelength of about 10 microns. Now, here's the curious thing: the atmosphere of the earth is selective in the radiation it allows through; it easily allows through the visible light from the sun, but it tends to trap the earth's long-wavelength radiation, reducing the amount that escapes into space. So if the atmosphere weren't there, we'd be colder than we presently are.

A lovely example of this is the common or garden greenhouse made of glass, or a motor car on a hot day with all the windows up. The glass allows the visible radiation to get through – the light from the sun – but it traps in the long-wavelength radiation that's trying to get back out again. So the temperature starts to build up until the glass itself gets hot and starts to radiate that energy out. And so the earth's atmosphere acts like the glass of a greenhouse: it stops the earth's natural radiation going back out into space and tends to keep more of it in than would otherwise be the case, but it allows very freely the sun's radiation to come through onto the earth. That makes us warmer than we would otherwise be and that makes life possible.

*KH: And the reason why the greenhouse effect is supposedly increasing, and we have global climate change, is because we have more carbon dioxide in the atmosphere, carbon dioxide that we've put in there and which makes the heat less able to escape.*

**PC:** That's right. If we ask 'What is this glass that covers the earth, or what is this atmosphere made of?', it's mostly made of oxygen and nitrogen. But as well there is water vapour, carbon dioxide and other trace gases in small amounts. The carbon dioxide and the water vapour are very effective in trapping that long-wavelength radiation from the earth and keeping us warm. Actually, the biggest greenhouse gas is water. It's by far the most important one. The carbon dioxide is the second most important, but water vapour is always there in the atmosphere as the oceans evaporate and rivers and lakes evaporate and we have precipitation – there's a natural balance.

*KH: So what is causing the change to that balance?*

**PC:** The view of most scientists is that it's the increasing carbon dioxide levels in the atmosphere. And anything that tends to increase the greenhouse effect may raise the temperature of the earth and cause more water vapour to go into the atmosphere, which could have the effect of even accelerating the rate of rise of temperature. So there's a real worry about global change in that once things start to warm up, there could be a runaway effect.

*KH: There seem to be so many variables, you see. Back to Michael Crichton for a moment – he says in the speech that I quoted from earlier 'To an outsider the most significant innovation in the global warming controversy is the overt reliance that is being placed on models'. And he says 'You can have catastrophic models for anything and the arrogance of the model-makers is breathtaking'.*

**PC:** I have again some sympathy with Michael Crichton on this – modelling is a very difficult business . . .

*KH: You put in this thing and you get this . . . worst-case scenario.*

**PC:** You have to be very careful. What are the parameters we're putting in? Do we really understand the earth's climate system well enough that we've got all the parameters at hand to put in so that the computer can do the calculations? Michael Crichton makes an analogy to the difficulty of weather forecasting.

*KH: So is the climate modelling any use?*

**PC:** Weather forecasting is a different business from climate forecasting. Climate is about the average behaviour while weather is about the fluctuations. There is a simple saying: 'Climate is what we expect and weather is what we get.' So we all know that weather is very complicated. Climate is more definite.

I have to say that within the climate modelling that has been carried out by many people throughout the world there's a lot of range in the predictions, but even at the lower end of the predictions it does look like we've got some serious temperature rises ahead of us.

*KH: What about the past? Wasn't there a medieval warm period? We didn't have carbon dioxide pumping into the atmosphere then . . .*

**PC:** Well, I think it's really interesting to look back historically. My own views about global climate change are not based on the climate predictions at all. Suppose we put the modelling to one side – what do we actually know and what have we actually measured? If you go back 2000 years and you look at things like tree-ring analysis or coral-ring analysis or whatever, the earth's climate – the earth's global temperature, the mean temperature – seems to have changed by about plus or minus 0.4 degrees. At the medieval period we're at the warmer time. It may have been the reason why Polynesian navigations and Vikings travelling over the oceans were easier. Then there was a colder period in about 1600 to 1700 – the Little Ice Age. Remember reading that people went out and had fairs on the frozen Thames? Then it got warmer again and that's where we are now. But what's happened in the last hundred years is a change of about 0.6 degrees upward. So this is a bigger change than we've seen for 2000 years, and it's all happened in a very short space of time.

*KH: So it's the size of the change that has convinced a large number of people that it's not just the earth's natural climate system coming and going?*

**PC:** Absolutely. And it's not just the size of the change; it's the rapidity of the change. It's happened really in the last hundred years and it's accelerated over the last 30 years and, furthermore, if you look at the temperature change and plot it on the same graph as the carbon dioxide increase – an increase that nobody argues about – the two follow each other. And I should point out that we know the carbon dioxide levels in the atmosphere back to 650,000 years ago now, and this information is obtained from drilling in Antarctica; drilling deep ice cores in Vostok at the geographic centre of Antarctica, we can see how much carbon dioxide there was in the atmosphere over a period from now back to 650,000 years ago. The carbon dioxide range is between about 200 to 280 parts per million. Now, in the last hundred years, we have increased from 280 to 380 parts per million. It's never been that high at any stage that we can measure back in the ice core history that we have, going back for nearly a million years. The other thing that we can tell from the ice cores is the earth's temperature, also going back 650,000 years. The way this is done is you measure the relative proportions of oxygen-18 and oxygen-16, which is in the ice. That's a very sensitive measure of what the temperature was at the time because the ratio of those isotopes at different levels of the atmosphere, particularly on the surface, depends very sensitively on the temperature. So we can get a pretty good idea of temperature variations as well, and we know by looking at these ice cores that every 100,000 years we've had an ice age, and then it swung around to a warm period, so we have a cycle.

KH: *So what's the cause of those cycles in temperature?*

PC: The explanation is that the earth's orbit slightly changes – in a completely natural way – because of the interaction of the planets. This is called the Milankovitch cycle. And where we are historically, now, is in a warm period. We should be heading, over the longer term, towards another ice age. That might give you cause for comfort. You might think 'Great, we're heading towards another ice age, global warming is nothing to worry about', but the time that it'll take to go down to a new ice age is about 50,000 years, and we're looking at temperature changes that have occurred over the last 50 years or so. So we're seeing rapid changes over a time-period that've never been seen over a period of about a million years. That's pretty compelling evidence . . .

*KH: And so the evidence of local cooling that people like Michael Crichton point to and say 'Look here, the glaciers are growing' is irrelevant?*

PC: Well, no. The glacier issue is very relevant and indeed there are some glaciers that have been growing. We New Zealanders know that the Franz Joseph glacier has been growing. I can remember when it was much shorter and in the last few years it's been growing again. But if you look at all the glacier information around the earth, and I've done this, glaciers have shortened by about a kilometre in the last hundred years – there's no argument about that. Franz Joseph may be growing but you have to look at all the glaciers and look at the whole pattern. The problem is that there may have been some heavy snowfall in the Southern Alps in a certain period, maybe 10 to 15 years ago, which is driving that particular glacier. So there are local reasons why glaciers may grow. But on average, over the whole earth, they are retreating.

I had a lovely experience recently at Chamonix. I was below Mont Blanc at a glacier called Mer de Glace, and I met this old lady, just standing there looking at the glacier coming down the mountain. I started to talk to her and she said, 'I've been coming here ever since I was a little girl.' She told me how the glacier had got smaller and receded over her entire lifetime. It was really quite moving – a personal story from someone who had seen, over her lifetime of 80-odd years, the glacier recession. Of course, as scientists we can't base our opinions on anecdotes like that. We've got to look at the whole pattern over the whole earth. And of course it's the same thing with temperatures – you'll find local regions where the temperature is actually decreasing.

*KH: Is there nothing in the way the world works that comforts you with the idea that it will correct itself?*

PC: Of course the earth will correct itself – the earth cannot be damaged by us humans. We cannot damage planet earth – planet earth will deal with this.

*KH: But it may wipe us out ...*

**PC:** It may wipe out humanity. And planet earth will go on for another 4000 million years.

**KH:** *I was hoping you would have a solution other than that, actually.*

**PC:** Let's face it – we humans have been around for less than a million of the 4000 million years on this earth.

**KH:** *This brings me to James Lovelock, someone with huge credibility as an environmentalist. He's now said, 'Time's up, civilisation is unlikely to survive; the temperature will rise 8°C in temperate zones and 5°C in tropics by the end of this century; billions will die and the few breeding pairs that survive will be in the Arctic where the climate remains tolerable.' That's like the bad news. The good news is you are one of a breeding pair; the bad news is that you have to do it in the Arctic. What do we think, Professor?*

**PC:** I think it's an extraordinary statement by James Lovelock but it's one that we should all take seriously, and certainly many in the science community do.

**KH:** *Lovelock's the man who came out with the Gaia idea decades ago. Let's talk about that because it's become part of the language. What does Gaia actually mean? It's quite a hard thing to get your head around.*

**PC:** Well, it's a way of looking at the earth as a living system. The idea is that the geology of the earth and the life-forms of the earth together evolve to produce conditions that make the climate and atmosphere amenable to life. So it's a way of thinking of the entire ecosystem of the earth, including the geology, including the climate, including the life, as one living system.

**KH:** *He describes it quite nicely. He says, 'We now see the air, the ocean and the soil as much more than a mirror environment for life; they're a part of life itself. Thus, the air is to life just as the fur is to a cat or the nest to a bird.'*

**PC:** It's beautiful isn't it? It's really quite poetic. The other idea is that humanity is almost like the mind and the spirit of Gaia and when we went out of the earth in the time of the moon landings and could look back and see the earth, suddenly Gaia could see itself. That's quite a poetic statement.

**KH:** *Teilhard de Chardin was saying something pretty similar long before this, and of course all the scientists went 'Oh, give me a break'.*

**PC:** They did. Absolutely. But actually I must say that many of Lovelock's ideas are pretty much mainstream today. I think the scientific evidence is there of this interdependence. There are obvious things such as plants helping to create the oxygen in the atmosphere. Before there were plants on the earth we had a carbon-dioxide-based atmosphere. But there are more subtle things. For example, there are little bacteria that live in the upper atmosphere that help seed clouds to make raindrops. So we know that living systems, often single-cellular living systems in their multiplicity, are part of the climate. In a sense, we as human beings depend on the whole biodiversity of the planet; we depend on the plankton and the bacteria and all forms of life to have an atmosphere and a climate and an environment where it's possible to live as humans. I think that's pretty much mainstream science now.

**KH:** *Just as a matter of interest, Richard Dawkins has said, 'How can you have planetary-scale homoeostasis evolve if natural selection works on individuals?' Can you see those two being compatible?*

**PC:** Yes, I can. I think Lovelock's argument is that the physical environment is very much part of the way in which evolution occurs and evolution of species determines the formation of the physical environment around us – through the atmosphere and through the effective atmosphere of the landforms – so I think that's a reasonable proposition. And I think that because Lovelock is someone who is a little bit ahead of his time, having come up with an idea that seemed pretty 'spacey' a few decades ago but which now seems to make some sense, he tends to be taken a bit more seriously now that he has made this dire prediction. I don't, however, think the prediction is based on any firm evidence that we have really reached the tipping point, but a feeling somehow on the part of a wise old man who has seen a lot in his life that maybe we have.

**KH:** *Well, yeah. The trouble is that it is so dire. In effect he is saying that it is too late to do anything about this.*

**PC:** I suspect he is wanting to make a wake-up call, and I think he has earned the right to do that.

**KH:** *But it's not – it's like a bail-out call.*

**PC:** But if you step back from what he's saying and ask the question 'Is it all over?', I think you might reasonably come to the conclusion that it's not necessarily so, and that there are things that you can do.

**KH:** *Nuclear power, he suggests, is one of them.*

**PC:** Well, for example. And of course James Lovelock's view on all of this is that the solutions that we all have in our hands cannot be in the form of some sort of a retreat into pre-industrial agrarianism such as mass organic farming. He sees technology and science as being central to the solution for the planet. And he does point out that, in the short term, nuclear power is an option *right now* that has minimum land usage and doesn't produce carbon dioxide . . .

**KH:** *It's interesting how everybody's got excited about this at this point because back in 1988 he took this position of pro-nuclear power in* The Ages of Gaia *saying, get over it because it's not a problem.*

**PC:** Yes, he has some quite extraordinary views on this. For example, he gives the example of this ecosystem that has developed around the Chernobyl power plant. The reason that people won't go there any more is because it's radioactive. But animals and birds and plants are colonising this area, and a whole ecosystem is developing there – oblivious to the radiation. He comes up with the rather extreme view that maybe the way to get more forestation and protect the planet from the developers and from the onslaught of humans is to spread a little bit of radioactive waste around. It certainly keeps people away and allows the ecosystems to grow again.

*KH: This is bonkers, right?*

PC: It's kind of outrageous, but it does get us to think. And I think what Lovelock's trying to tell us is to avoid cosy, sentimental mysticism about this problem: we've got to look at this thing in a very fresh way – we've got to look at this in a brutally honest way, and we've got to use science and technology to solve the problem. For example, we have to find transport fuels and electricity generation schemes that are carbon neutral at least. Biofuels almost certainly have a place, especially in New Zealand. Wind farms are helpful, but not if they increase pressure on land use.

*KH: Let's just go back to his prediction about the temperature rising 8°C in temperate regions and 5°C in the tropics by the end of the century. Now, this is big. Is this associated with the 'tipping point'?*

PC: Yes, Lovelock and other scientists do refer to the 'tipping point'. The tipping point is the non-linearity in a system – it's where the response of the system is out of all proportion to the stimulus or the provocation. For example, if you lean back on a chair, a little further and further, suddenly you tip and you're in trouble. If you pull a rubber band and it stretches and stretches, suddenly ping, it's gone . . .

*KH: So all of a sudden you get some sort of a chain reaction going on?*

PC: Yes. And the problem is that although we've only had a 0.6°C temperature rise this century, the rise has been far more rapid than we've seen in 2000 years. We suspect it's probably anthropogenic: it's human-related, the carbon dioxide that's been pumped into the atmosphere. Now 0.6°C doesn't sound like very much when you talk about mean temperatures – after all we're used to having temperatures going up 10°C from one day to the next – but mean temperatures are hugely important. After all, the previous ice age 100,000 years ago was only 5°C average temperature lower than it is now. So this mean temperature rise seems small but the problem is that it could lead to some non-linear positive feedback effects.

*KH: Such as?*

**PC:** I'll give you some examples: the first is that as we see the polar ice caps and the sea ice starting to shrink in area, and in the Arctic we've seen 300,000 square kilometres reduction in 30 years, then the *albedo* of the earth – the reflectivity of the earth – starts to change.

*KH: What did you say?*

**PC:** It's called the 'albedo' – it's how white the earth looks from space.

*KH: Really? Albedo.*

**PC:** Yes, how bright we are. So a lot of the brightness of the earth is due to the reflection of the ice caps. As those ice caps start to retreat we reflect less sunlight and therefore we absorb more and we become warmer, so that's the sudden positive-feedback effect. As the tundra starts to melt it releases methane into the atmosphere and methane is quite a deadly greenhouse gas – with great global warming potential.

*KH: Why does it release the methane?*

**PC:** Because the methane from rotting vegetation in the tundra is sealed off by the frozen surface, and once that starts to melt there's a release. The oceans, instead of becoming a net sink of carbon dioxide, start to become a source of carbon dioxide if they're warmed up too much. We've already had a warming of the surface. Warming of the surface can affect the nutrient flow to the plankton, which is another absorber of carbon dioxide, so if we do anything to damage the plankton growth, we stop the ability of the oceans to sequester carbon dioxide. And if the surface temperature of the earth rises, we change the balance between respiration of the organisms in the soil to photosynthesis. So what we worry about is whether with a certain increase in temperature, suddenly we can trigger a massive release of carbon dioxide, and that could lead to massive global warming with all the catastrophes that Lovelock talked about.

*KH: And Lovelock says we have reached the tipping point?*

**PC:** The truth is that we don't know if we've reached the tipping point – there's no evidence that we have. But the point is that there's a possibility. And here's the thing: it's called risk management. If you're a parent and your child is climbing up the railing on the Interisland Ferry, there's a possibility the child might tip over; there's no proof but, as a parent, you aren't going to take that risk because the consequences are absolutely too terrible. And it's a bit like that on earth – we are facing an uncertain risk, but with consequences that are catastrophic. And I think that's the challenge for the policy-makers, for all of us.

*KH: And because, as Lovelock said, when catastrophe happens and some of us are reduced to breeding in the Arctic with somebody we've probably never met before, people are going to say, 'Why did those Greens stop us from having nuclear power?'*

**PC:** They might, and I think I would agree at least with Lovelock's argument that we have to look to science and technology to solve a lot of the problems – a retreat to pre-industrial lifestyles is not going to work for us, not with 6.5 billion people on this planet.

*KH: Well, apart from being depressing, it's fascinating.*

# CHAPTER 15

# THE COSMOS

The most exciting phrase to hear in science, the one that heralds
new discoveries, is not 'Eureka!', but 'That's funny . . .'
– ISAAC ASIMOV

KH: *All right, then, Paul. The Big Bang – is it true?*

PC: It's an amazing story. There seems to be a lot of evidence for it, and the more we look into the universe, the more the Big Bang makes sense. It seems to be the way our universe started.

KH: *Let's recap. The universe is 100 billion – oh, my brain curls up – galaxies, right?*

PC: Ah, well it's big!

KH: *Oh, you're disputing that?*

PC: No, there are heaps of galaxies.

KH: *Each with about 100 billion stars? Are you saying that's an exaggeration?*

PC: No, what I am saying is that there is a universe out there that we can see. That's the observable universe. And, of course, it takes time for light to get to us. And as far as we can see out, it's taken about 10 billion years for that light to get to us. So when we look to the very edge of the observable universe, we are seeing it as it was 10 billion years ago. But there may be more universe beyond that which we can't see because there hasn't been time for its light to get to us. So, we have to distinguish between the total universe and the observable universe. And, as a result, we don't really know how big the universe is.

*KH: No. So how do we know about the Big Bang, or think we know about the Big Bang?*

PC: Well, the first indications were the measurements by Edwin Hubble in the 1930s when he discovered that the more distant a galaxy was, the faster it was appearing to move away from us, here on earth. He could tell how fast stars in a galaxy were moving away by looking at the spectrum of the starlight. It is possible to look at the detailed information in the pattern of colours and identify light from hydrogen atoms, so Hubble could say 'Look, there's hydrogen, but with all the colours shifted to slightly reddish'. And that red shift is something that would happen if a star were moving away, something called the Doppler effect, and one which we are all familiar with when we hear the drop in pitch of a car horn when the car moves away from us. And so what Hubble discovered was that the more distant a galaxy and the more distant its stars, the more red-shifted it was, and the faster the movement away. And that suggested an expanding universe. By extrapolating back in time, the expansion suggests a moment when all the matter in the universe was together. Georges Lemaître was the first person to draw that conclusion.

*KH: Hubble also proved that the Milky Way was not the entire universe, didn't he?*

PC: Yes, there were issues back in the 1930s about the nebulae seen through telescopes, and whether these were separate galaxies or whether they were within the Milky Way galaxy.

*KH: Just a sceptical interruption here. Aren't there other explanations for the red shift than that galaxies are receding?*

PC: There was a physicist called Fritz Zwicky who had a 'tired light' theory – that light lost some energy escaping from the gravity of stars.

*KH: Tired light. I like that!*

**PC:** But it didn't really stack up in the end. And there is so much else which supports the idea of an expanding universe. In fact, if you go back to Einstein's cosmological equations, there has to be something like an expansion of space to make those equations work: otherwise, with the matter that there is in the universe, the gravitational attraction would cause the universe to collapse in on itself. Einstein was always aware of that and he put a kind of fudge factor in there, one he called the cosmological constant. But now we know that the expanding universe balances the attractive forces due to the matter and energy density. And that's not the only evidence for the Big Bang, or that the age of the universe is, as we think, something like 13 billion years.

*KH: So what else?*

**PC:** There's evidence such as radio emissions from space, radio signals that would be associated with the early stages of the universe. We find that these emissions come from the very outer part of the universe that we can see. In other words, only the very early universe produces those radio waves, which is consistent with the Big Bang idea. But probably one of the most convincing pieces of evidence is to do with how much helium there is compared with how much hydrogen. If there had been some fantastic explosion at the beginning of the universe that generated all the different atomic nuclei available to subsequently form stars, then as the expansion occurred in the furnace of the Big Bang, the hydrogen nuclei which were fusing together to make helium would eventually be too far apart and the helium generation would close down. We would then be left with a particular imprint, the ratio of helium to hydrogen, of about $1:4$. That imprint would be an indicator of a stupendous explosion at the beginning. And when we look at stars, that $1:4$ ratio is exactly what we find.

*KH: All of the stars?*

**PC:** All of the stars. And in fact if we were to find stars that didn't have that ratio, then that would be contrary to the Big Bang theory.

**KH:** *So that would be, as it were, the Precambrian rabbit fossil!*

**PC:** Yes, this would be the cosmological equivalent of finding J.B.S. Haldane's Precambrian rabbit fossil. Such a fossil would contradict the theory of evolution. And a wrong helium to hydrogen ratio would contradict the Big Bang. Every good scientific theory has some test, some observation which, if made, would disprove it. But there is other important evidence as well. And probably the most famous is the microwave background radiation from space.

**KH:** *This is the cosmic microwave background, CMB?*

**PC:** The cosmic microwave background. That's it. It was discovered in 1964, quite by accident, by Arno Penzias and Robert Wilson. They were engineers, looking at how to use microwaves for communication purposes. They found that there was an annoying 'hiss' in the background, and the source of this seemed to be from the sky or beyond, and when they aimed their antenna at different parts of the sky, it seemed to come uniformly from all directions in space. When Penzias and Wilson analysed the microwave frequencies associated with the 'hiss' they found a remarkable pattern. This pattern seemed to be exactly what one would find from natural thermal radiation coming from something at thermal equilibrium, at the temperature of 3 Kelvin (about −270°C). At that very cold temperature, thermal emissions are right in the microwave part of the electromagnetic spectrum (by contrast with objects that glow 'red hot', that is with visible light, when they are very hot, say 1000 Kelvin). When the astrophysicists heard of Penzias and Wilson's results they were very excited, because the Big Bang theory predicts that there would be a radiation residue of that original explosion and that, because of expansion of the universe since the Big Bang, that radiation would be at exactly the microwave pattern corresponding to about 3 Kelvin. The cosmic microwave background is another lasting imprint of the Big Bang.

**KH:** *And that's the definitive proof is it? So far?*

**PC:** Well, every little bit of extra evidence that is found adds credence to the theory. I don't think you ever finally prove a scientific theory, in the way that it is indeed possible to disprove a theory. But when you get a weight of evidence, when all these different sorts of observations start to click together, then it starts to look pretty convincing.

*KH: So, the universe is expanding, but it's more correct to say that space itself is expanding?*

**PC:** Yes, it's space itself that is expanding, and that's a very curious notion.

*KH: Yeah. The only way I can handle that is through the balloon idea.*

**PC:** Exactly. If you think about living on the surface of a balloon, when the balloon is expanding, then your world on the surface of that balloon is also expanding. Parts of that world are getting further and further apart. Of course, that world would be two-dimensional, the world of the expanding surface. But we live in a three-dimensional space which is expanding, and this is much harder to visualise. In truth, we live in a four-dimensional space–time. One of the things that Einstein taught us through general relativity is that space and time are intimately connected. So if you go back to the origins of the Big Bang, when space was infinitesimally small, then that's also the origin of time.

*KH: So the Big Bang was not an explosion in space, but an explosion of space.*

**PC:** *Of* space. And so you can't really answer a question like 'what was before the Big Bang?' There is a singularity at the origin which means there is no meaning for time before that moment.

*KH: And a singularity is zero size and infinite density?*

**PC:** Absolutely.

*KH: Which doesn't make any sense at all, Paul.*

**PC:** It's a very hard idea to get your head around, isn't it?

*KH: But my brain still requires to know what was before the Big Bang.*

**PC:** I think that's what everyone's question would be. And it's even worse than that. There's a short period just after time zero that physicists can't deal with, and this is called the Planck time. There is a time – it's actually $10^{-43}$ seconds – below which physics makes no sense. It makes no sense because of the Heisenberg Uncertainty Principle. In order for such a short time to have meaning, you would have to have a very large energy to measure it, and that energy would be so compact, it would collapse by gravitation into a black hole. Hence the physics wouldn't be describable at such short times. What this means is that we can't even describe that instant just after the Big Bang. The physics is too complex there.

*KH: That seems to be a bit of a cop-out.*

**PC:** Well, it's not really a cop-out. It just means that there remain questions to be answered. And that's great in science, to have things that are as yet unknown and as yet not understood.

*KH: I'll tell you what really worries me, and that's this whole idea of dark matter. Dark matter is required because you need more gravitational pull out there than visible matter would account for.*

**PC:** That's right. And even if we look at the rates at which galaxies are rotating, for numbers of stars that we can see, there's not enough mass to account for those rotation rates.

*KH: But doesn't that seem dodgy to you? I mean, dark matter is something we've invented because what we believe doesn't make sense unless we invent something else. That's wrong.*

**PC:** You sound a bit like Ernst Mach when you say that.

*KH: Do I?*

**PC:** Well, Mach said that Boltzmann had invented atoms to explain thermodynamics and that we can't see these atoms. They're terrible things. They don't exist! The point is, you model this. And, as for atoms where Boltzmann was vindicated, later observations may find dark matter, so verifying that model. Or later observations could disprove the model. That's the way that science works.

*KH: Dark matter seems such an unlikely thing to invent, though.*

**PC:** It is hairy and there is a whole lot of uncertainty around this idea – of how much energy there is in the universe, of how much matter there is. But one thing we do know is that the amount of matter and energy does seem to closely balance the expansion rate. In other words, that the universe expansion is 'flat'. It looks like the expansion will continue indefinitely, but slowing all the time. And that in itself says something about the way the Big Bang must have occurred, that flatness of the expansion of space.

*KH: What do you mean 'flatness of expansion'?*

**PC:** What it means is that there is almost a perfect balance between the rate at which the universe is expanding and the amount of matter and energy that's trying to contract it by gravitation. A balance in the competition between these two things.

*KH: Handy, then?*

**PC:** That's the way it seems to be. A universe continuing to expand at a diminishing rate of expansion.

*KH: So no 'big squeeze', then?*

**PC:** Most cosmologists would say not. If there was to be a big squeeze, it's unusual that it hasn't already happened. Once you go out of this near-perfect balance that I am talking about, things get out of hand quite quickly. And the quite remarkable balance we have must have its origins in the Big Bang itself. The cosmologist Alan Guth has given us an idea which could explain that balance. It's known as 'inflation', a very early period (below a fraction of a second) after the Big Bang, during which there was a period of rapid acceleration of expansion. Why? No one really knows for certain. But it would have had the effect of producing the very flatness that we have been talking about.

*KH: So what does this mean for life in the universe?*

**PC:** Well, here we are in the universe, and if we look at the way the universe seems to be structured, it seems to allow life to occur. There is something quite remarkable about that. There are many aspects of the way that physical laws have emerged in the universe which seems to make life possible.

*KH: Is this something to do with the 'anthropic principle'?*

**PC:** Let me try to explain that. Some people looking at the universe conveniently allowing life to emerge might be tempted to think that there is some grand design. But suppose I bought a Lotto ticket and I watched the balls coming down the tube to produce the winning number, my number. I might think there was some real designer purpose behind the fact that those balls shook out exactly right for me. But I would be forgetting about the million other people who didn't win. Or suppose I was a soldier surviving a terrible battle in which most of my comrades were killed, I might think there was some divine purpose or special conditions favouring me. Anyone who was the survivor could think that. But of course the conditions were special for that survivor. Simply by being the survivor, one is able to see the special fortune that led to survival. The anthropic principle is a bit like that. Of course the universe is right for life. After all, here we are, alive in the universe. So the conditions must have been right. So we think that it was designed that way.

*KH: So that's the anthropic principle?*

**PC:** Almost. Of course our universe sustains life, because here we are. But what if there were many other universes that didn't sustain life? Then only in the one that did sustain life could we ask this question. It may be that our universe is just one of many universes in a 'multiverse'. Others are not right for life. Ours obviously is. That's the anthropic principle.

*KH: This may seeem a very stupid thing to say, but it never occurred to me that it is called 'uni'-verse because it is only one.*

**PC:** I guess words have a certain power to guide the way we think. But the idea of a multiverse is curious. And often physicists tend to dream up ways of solving very difficult problems, like this anthropic principle. There is a related case in quantum mechanics where the outcome of a measurement can be any one of many possibilities. And one explanation is that all possibilities do in fact occur, but in different universes. That's known as the 'many worlds' interpretation of quantum mechanics. The theoretical physicist John Wheeler said of this idea that it was 'cheap on assumptions and expensive on universes'.

*KH: I like that response. On another matter to do with cosmology, what's Dirac's coincidence?*

**PC:** Paul Dirac discovered a really interesting number, that is, $10^{36}$.

*KH: Ah!*

**PC:** That number is how much bigger than the gravitational force the electromagnetic force is. And the coincidence is that $10^{36}$ is also how much bigger the observable universe is, compared with the size of the smallest particle that makes up matter, which is the proton.

*KH: Hold on. Say that last bit again. What's this $10^{36}$?*

**PC:** Well, $10^{36}$ is the ratio of the size of the electromagnetic force, the thing that binds atoms together, compared with the size of gravity, which binds stars together. Gravity is very much weaker than electromagnetism. But $10^{36}$ is also that ratio of the size of the observable universe and the size of a proton.

*KH: But if the universe is expanding, isn't the size of the observable universe growing?*

**PC:** Yes, but at this time, at this epoch, we know how big it is and this is what Dirac's number is based on. But there turns out to be a simple explanation for Dirac's coincidence and it's to do with the lifetime of a star. Our epoch, the point in time since the Big Bang when we measure the size of the observable universe, is roughly (that is, give or take a factor of two or so) the lifetime of a typical star after the Big Bang. And the lifetime of a star is determined by how long it takes a photon, which carries the heat, to get from the core of a star to the outside, and that time depends on the star's size, and that size, the size needed to make a star 'burn', in turn depends on the ratio of the gravitational force to the electromagnetic force. This so-called coincidence is not a coincidence at all. It is a natural consequence of the physics, once the universe gives us the number $10^{36}$ as the ratio of the gravitational force to the electromagnetic force. That special number determines how long stars live, a pretty important consideration when you are thinking about how long it takes life to evolve. In fact there are a whole lot of special numbers in the way our universe operates.

*KH: Hang on. I'm just pre-empting all those people who would say 'Ah yes, scientists call it a coincidence, but they are nasty godless creatures'.*

**PC:** Well, let me tell you about another one of these special numbers. It's associated with something called the ripples in the universe. The Nobel prize for physics in 2006 was won by two physicists, George Smoot and John Mather, who discovered the ripples in the microwave background radiation using satellite imagery. Those ripples are about 1 in 100,000, a very subtle variation, and one which happens to be the same as the variation in the mass density of the universe. You won't be surprised if I tell you that this equivalence is predicted by the Big Bang theory! Anyway, that mass density variation of 1 in 100,000 is what enables stars to form and galaxies to form. If the ripples were bigger there would be mostly black holes and few stars; and if smaller, then stars wouldn't be able to form at all. Either way, the variation would preclude life.

*KH:* Is that one of Martin Rees's six numbers?

**PC:** Yes, it is – another remarkable number that just enables the right conditions, the right pattern of stars and planets for life to emerge. No one really understands why the universe is so fine-tuned with exactly the right numbers to enable life to occur.

*KH:* So what determined these numbers?

**PC:** They must have been set at the earliest instance of the Big Bang. That aspect is mysterious.

*KH:* To have exactly the right numbers? The right mix of elements?

**PC:** Yes. And, in another sense of the word 'element', the right mix of elements to enable the chemistry that we see, the chemistry that supports life. The role of carbon in particular. There is a remarkable aspect to carbon; the fact that it forms at all is related to a very strange quirk of nuclear physics. Protons fuse to make helium, which fuse to make beryllium, but the next step, to get helium and beryllium to fuse to make carbon, should actually be quite difficult. There happens to be an 'accident' in the way the nuclear physics of the reaction occurs. It needs something called a resonance.

*KH:* Is that Hoyle's insight about carbon?

**PC:** Yes, Fred Hoyle realised that there had to be this effect to make the helium–beryllium reaction work, and he suggested to the nuclear physicist Fowler that he look for a resonance at an energy of 7.65 million electron volts.

*KH:* And Fowler said, 'Don't be stupid, I've got more important things to do'?

**PC:** I think he said, 'I've looked before and there's nothing there.' But he looked again more carefully and found it. Fowler got the Nobel prize and Hoyle missed out, but that's another story.

*KH: So coming back to Martin Rees's book* Just Six Numbers, *he says there are six numbers which determine the conditions in the universe, which make it as it is, which determine the conditions that make life possible on this planet.*

**PC:** Well, that's his choice. I think you could go on and on here and find more numbers that make life possible.

*KH: But his point is 'Look, how exactly right these six things had to be. If they were slightly weaker or slightly stronger, then everything would be different.' So you can understand why some people would say, in the words of William Blake, that such 'fearful symmetry' requires the hand of a creator.*

**PC:** I can understand that. But it's not something that is satisfying from a scientific point of view, because such a position just poses a new question – what made the creator? It replaces one question with another question.

*KH: Let's not revisit that, but . . .*

**PC:** I can undertand why people see the hand of the 'designer'. It's remarkable that we live in a universe that makes life possible. But then, of course we would say that, if you go back to the anthropic principle. I mean, here we are, you and I are having this conversation. Hence, it *must* be that our universe makes life possible.

*KH: The issue is whether it's good luck or good management. Isn't it?*

**PC:** It would be luck if there were multiple universes, only one of which had conditions that favoured life. We are clearly in the one that allows life. We are like the person who won Lotto and saw the hand of the designer in that. We can't forget the losing cases, if indeed there is a multiverse. Of course the conditions are right. We are here. That's the anthropic principle.

*KH: So what about life elsewhere in our universe?*

**PC:** Here's another fascinating thought. It's taken life on earth, at least the form of life which can ask questions like the ones we are discussing, about three billion years to evolve on earth. And that's roughly the lifetime of a star. Our star will go on for a few more billion years but, again, give or take a factor of two or so, evolution has taken about a star lifetime. We are about halfway through the sun's life. Why is it that life on this planet has taken about the same time to evolve as the age of a star? If our planet were fast in that sense, and if it took much longer elsewhere in the universe, then there is not much chance of any life evolving far enough to have the conversation we are having, before the star supporting it burned out. But suppose, on the other hand, life on earth was slow, that elsewhere intelligent life evolved much more rapidly than the lifetime of a star. Then there would have to be many, many cases of intelligent life evolving elsewhere in the universe. And that poses the question, why haven't they communicated with us? Why haven't we heard from them? So there is another conundrum.

**KH:** *Do you personally think that there is life elsewhere in the universe?*

**PC:** I think it is very likely.

**KH:** *More than unicellular? I mean, intelligent life?*

**PC:** Yes, I do mean life that could ask the sort of questions we are asking. I do think it is likely. But when you think about the likely lifetime of a civilisation or period of time over which intelligent life could develop before its demise, then the chances that their communications would reach us, during our period of civilisation, before human demise, become very small indeed. Life may be out there, but whether we can communicate with it is another question again. Most unlikely given the vast scale of the universe. In that sense, I think that we are alone.

CHAPTER 16

# QUANTUM MECHANICS

Quantum physics is the most successful and utterly
bizarre theory in the whole of physics.
– SIMON SINGH

**KH:** *I've just been reading Simon Singh.*

**PC:** Oh yes, wonderful writer.

**KH:** *And Simon Singh quotes St Augustine, who, in his confessions, answers the question of what God was doing before he created the universe, with – and this is St Augustine in AD 400 – 'Before He created Heaven and Earth, God created Hell to be used by people such as yourself, who ask this kind of question', which I like very much. Anyway when somebody says 'quantum mechanics' or 'quantum physics', what do they mean?*

**PC:** Well, it's a particular theory within physics that enables us to explain the behaviour of atoms and photons and electrons and molecules, but it actually applies to absolutely everything. It applies to large things like Paul Callaghans as well; so, essentially, if quantum mechanics is true it should really apply to big things as well as small things.

**KH:** *Well, Simon Singh's book is called* Big Bang *and he says that's 'quantum cosmology'.*

**PC:** It is indeed, and at the very beginnings of the origins of the universe – at the time of the Big Bang – quantum mechanics was playing a major role because everything was very compressed and very small. Later on, it's not so apparent perhaps, but in truth, it lies at the heart of everything, from how stars shine to how they collapse into white dwarfs or neutron stars. But we see quantum mechanics working very explicitly when we look into atoms or look into the regime of the very small.

*KH:* *So what is at the heart of quantum mechanics?*

*PC:* Physicists like to think of a universe comprising waves and particles. Cricket balls and Callaghans are like particles. They have mass. We can say where they are at any time. On the other hand, radio transmissions and the ocean surface are examples of waves. They are separate from the medium that carries them. They move, they carry energy but they have no mass, and they are everywhere, not in a particular place. One key idea of quantum mechanics is that particles can be both waves and particles at the same time, and that the 'wavelength', the distance from crest to crest of the wave, is inversely proportional to the particle momentum. We don't notice wavelike effects with heavy objects, like Callaghans or cricket balls, because the wavelengths are so short. But for very light objects, like atoms or electrons, where the wavelengths are long, the effects are extraordinary.

*KH:* *So, for example, with electrons, according to Rutherford's original model of the atom, it was assumed that electrons orbited around the nucleus, but you are saying they don't?*

*PC:* Well, no. Rutherford's model is basically right. There is a nucleus and there are electrons 'orbiting', but if we think of the electron as being like a wave, it means it's everywhere in the atom, at all times.

*KH:* *But it's also a particle. It's a wave and a particle.*

*PC:* And this is the deep mystery behind quantum mechanics. We're all familiar with waves. We look at the ocean and we see the waves coming in, and if you were to say 'Where is the wave?', well, the wave is sort of everywhere. It's all along the beach; it's not in one place. So waves are things that are spread out in space.

*KH:* *But at any particular moment in time, though, that wave is at a particular point.*

*PC:* Well, it isn't. Because the wave, as it comes in, is linked to crests way out to sea, and it is stretched right along the beach, so that the surfer travels up and down one crest of the wave. So waves are things that are spread out. And that's the very strange thing about quantum mechanics because if a particle like an electron can be a wave, how can it also be a particle? That's something deeply troubling. It's sort of an insanity at the heart of quantum mechanics. There's something about quantum mechanics that is infuriating, because it's so strange and yet it seems to work and it seems to give the right answers to every measurement we make. But a lot of people hated it – like Einstein, for example.

*KH: But didn't Einstein say that light behaved like particles?*

**PC:** He did. He explained something called the 'photoelectric effect' this way. So quantum mechanics works both ways, imparting wave-like properties to particles and particle-like properties to waves. Einstein was part of that early understanding, but he parted company with it later on.

*KH: So is this idea of waves and particles connected to Heisenberg's 'uncertainty principle'? Can you explain that?*

**PC:** Well, it's a very strange principle because it says that there's only so much we can know about nature. We can know some things very accurately but if we know that accurately, then other things we can't know well. And a good example of that, going back to our electron, is the question 'Where is an electron?' If we want to know the answer to that question with some accuracy, then we can't know the answer to 'How fast is it moving?' Or, alternatively, we can ask 'How fast is it moving?', but then we can't know exactly where it is. Now, that's not something we're used to. When we describe a cricket ball we are used to saying that it is at a certain place in space and it has a certain speed. In the normal world of everyday life we're used to being able to describe things accurately.

*KH: Remind me about why we can't know where it is and how fast it's moving at the same time.*

**PC:** Well, it's because there's a particular scale in nature that is defined by something called 'Planck's constant', and this means that when we get to things that are very small, the physics starts to get different. You can think of this very easily with an electron this way: how would you know where an electron is? Well, you could shine some light at it. You could bounce one of Einstein's photons off it to find out where it was. And how would we find out where it was? We could take the light through a lens and focus it on a point. Use a microscope, in other words. Now, to look at things very accurately with a microscope you need a big lens. The bigger the lens, the more accurate. But the bigger the lens, the more uncertain is the path that the photon took to go through the lens. Did it go through the right-hand side of the lens or the left-hand side of the lens? And if it went through the right-hand side or the left-hand side, it would've kicked the electron slightly differently. Because to look at that electron we have to scatter the photon off it and the photon has to slightly hit that electron to come off, to get into our eyes. And that little kick gives some uncertainty to the velocity of the electron momentum. So the bigger that lens, the greater the amount of kick that's possible to the left or right, and the bigger uncertainty there is about the velocity.

So that's the way Heisenberg saw it. But Niels Bohr saw it even more fundamentally. He said, 'No, this is just a feature of nature itself. It's not just how we can do a measurement; it's that nature forbids us to know both the position and the momentum with absolute precision.' That's a very strange idea. And it's something that Einstein really hated. He said, 'Look, nature is there – there's a reality. Maybe we don't know how to describe it, but there is a truth there.'

*KH: Yeah, he wanted the cosmos to be unchanging as well. Didn't he?*

PC: He did, yes.

*KH: He tried his best to fix that.*

PC: He did and he went astray in all sorts of ways. Really, I suppose, at the end of his life he was somewhat cut off from the mainstream of physics.

*KH: He knew, he kind of knew emotionally, didn't he?*

**PC:** Well, I think that he felt *instinctively*, I would say, or rather guess. In a sense Einstein believed he had a view of the universe which was deep and profound. But we now know that in many ways he was wrong.

*KH: Talking about Niels Bohr, Simon Singh quotes Niels Bohr as saying that 'anyone who is not shocked by quantum theory has not understood it'.*

**PC:** I think that's absolutely true. I think Richard Feynman said the same thing – 'there is no one who understands quantum theory'. And I think that's very true. I teach my students that, and I say, 'Look, here's how quantum theory works, here are the tools to calculate it, but we really don't understand *why* it works.'

*KH: No, it's quite mad, isn't it?*

**PC:** It is quite mad, but it is probably one of the most successful theories of physics that ever was.

*KH: And that's an interesting thing in itself because what you're saying is, it's successful because everything seems to fit.*

**PC:** Everything seems to fit, yes.

*KH: But it will cease to be successful when one thing doesn't fit.*

**PC:** If one thing fails – if it fails one test – it's going to be in serious trouble. This is the tough thing about physical laws: to be a law or a successful theory, it's got to work every time.

*KH: Coming back to the electron, what's this business about how you have to travel twice around it to get back to where you started? Go back to basics. An electron is . . .*

**PC:** An electron is a point-like particle which carries negative charge and has mass. It was discovered by J.J. Thomson in 1897.

**KH:** *Negative charge?*

**PC:** Negative charge. And it's the basis of all electrical things that we know. All of electricity, all of electronics – it's what carries the charge.

**KH:** *Well, what about the proton? The proton's got the positive charge.*

**PC:** The proton is there stuck in the nucleus of the atom. It's not as mobile as the little electron that can dash around and do all sorts of useful things.

**KH:** *All right, so here's your atom, and stuck in the nucleus is the proton with the positive charge and the neutron with no charge, and around the outside of it are the electrons.*

**PC:** Right. Let's go back to the story of Rutherford's atom and the spectrum of light coming out of the atom. If we look at a candle flame and we pass that light through a prism we see there's lots of separate colours in there, and those colours are associated with the 'quantum jumps', jumps that involve electrons changing their orbits in the atom. But what makes those particular orbits or energy levels in the atom? It was a physicist called Erwin Schrödinger who realised that you could explain all those particular orbits that the electron had in the atom by thinking of the electron not as a particle, but as a wave.

Now we're used to the idea of particular states for waves. If you pluck a guitar string you have a note. That note has a particular timbre. It might be middle C, but it sounds quite different to middle C on a trumpet, or middle C on a flute, and that's because of the harmonics – the arpeggio of notes that are present when you pluck the guitar string – the octaves, the perfect fifths, the major thirds and so forth.

**KH:** *So what are you saying it's got? It's got a sound and a note?*

**PC:** That's right.

**KH:** *And they're quite different?*

**PC:** They're quite different. And it's those harmonics on the guitar string that gives it its particular character. A guitar string has a fixed length and it confines the wave to be 'standing', and that's what causes the harmonics. So if an electron is a wave and is confined in an atom, it'll have harmonics. There will be a 'timbre to the note', if you like, of an electron in an atom. If it's a wave it's confined in a region of space – the way that a wave in a guitar string is confined to the particular length of that guitar string. And those harmonics of the electron exactly explain the spectra of atoms, so that was a remarkable achievement.

But there was one very strange feature. The spacing of the 'notes' should have been integral amounts of a unit of rotation of the electron (known as Planck's constant). But when physicist Wolfgang Pauli looked very closely at those spectra he found half-units there.

**KH:** *What do you mean?*

**PC:** Well, quantum mechanics says that the waves are standing in the atom in the form of harmonics, then you have certain rules about what the spacing of the notes will be – and those rules were violated by having half-intervals. Those half-intervals weren't actually predicted. But they were there – you could see them in the spectrum. Now, if you had these half-intervals, it meant the electron's behaviour was very strange. One of the strangest things that resulted was that if you went around the electron, you would need to go around twice to come back to where you started. If you went around once, that wouldn't be enough, and that's very strange.

Actually it's not something that is entirely unfamiliar to us. If I take a piece of paper, a narrow strip of paper like a ribbon, and if I made a bracelet out of that paper by turning it around and joining the ends together...

**KH:** *You made a circle.*

**PC:** I made a circle. And if I put my finger on the outside and went around that, I could retrace my position right back to the beginning in one orbit around the bracelet. But suppose I made a twist in the bracelet before I joined it up, so instead of joining the ends together I put a twist first. Now let's do the same trick. I start on the outside, I go around, and then I come to the twist and that takes me underneath. That's not where I started. I have to go around a second time to get me back to where I started.

*KH: OK. Is that a metaphor?*

**PC:** It's a metaphor, and it's a mathematical metaphor. We're not unfamiliar with the idea that there are certain objects where you have to go around twice in order to come back to where you started. So the electron just happens to have this feature.

*KH: How do you know?*

**PC:** Well, we know from looking at the spectra of atoms – that's the first clue to this. But then you can start to measure behaviours of electrons: you can start adding electrons together to see if they combine in such a way that shows this feature. And they do. And one of the features that happens when you combine electrons is known as the 'Pauli exclusion principle': that no two electrons can occupy the same state.

*KH: What do you mean the same state?*

**PC:** The same wave, if you like. The same wave harmonics, because their waves would cancel each other. And that's why materials are hard. If I pick up a little pebble, for example, and squeeze it, it feels hard. But Rutherford's atom, as we call it, is full of empty space. Why is it hard? Why is it when atoms are all close together – as they are in the little pebble that I'm holding – that when I squeeze it I can't just push into that empty space and make all the electrons crush down into the nucleus.

*KH: What's the answer?*

**PC:** The answer is that the electrons absolutely will not occupy the same states, and when we try to squeeze that pebble, we're pushing on the electrons and making them come closer together and trying to force them into the same state.

*KH:* *And why won't they occupy the same state?*

**PC:** It's because of this peculiar symmetry that's associated with the fact that you have to go around twice to get back to where you started that forbids that, and that's known as 'anti-symmetry'.

*KH:* *What is that? What is 'anti-symmetry'?*

**PC:** It's the need to go round twice, to come back to the start – the effect that I demonstrated with the little piece of paper called the Möbius strip. It can also be described as the sense of 'up-ness' and 'down-ness'. This 'anti-symmetry' property of the electron, which physicists call 'spin', is absolutely at the basis of everything that we know in the universe. It explains so much. It explains all of chemistry – even things like living processes such as photosynthesis. Understanding how that works comes back to this spin property of the electron: the strange anti-symmetry of the electron that we've been talking about. It reveals itself in so many ways. And I'll give you a wonderful example of a technology where we depend upon that: magnetic resonance imaging. There, it turns out the proton has the same property: spin. And the protons in our water molecules when placed in a magnetic field can produce radio messages, which we use to create wonderful images inside the human body. So that's a technology that completely depends upon it. But really there's practically nothing around us in the physical world we can explain without going to this idea that the electrons have the strange anti-symmetry that we've been speaking of.

*KH:* *The reason it's called 'quantum mechanics', as I understand it, is because Niels Bohr, when he worked with Rutherford, came up with the idea – or the fact or the theories – that the electrons can move from one place to another instantaneously without going in between. And that's the quantum leap?*

**PC:** That's right, though that's what we call the old quantum theory. That's where the beginnings of the quantum theory came from. You had the lumps, you had the jumps –

*KH: Do we not believe that any more?*

PC: No, no, we do, we know it's true, but it didn't tell us much about the why. It didn't tell us why we had these particular states that electrons could jump between. And it wasn't until Schrödinger taught us to properly describe the electron as a wave that we could understand. Then we could see it being like the vibrating of a guitar string or the modes of a drum or tympani . . .

*KH: Have you already answered my question, then, about how an electron can move from one place to another instantaneously without going in between?*

PC: No, I haven't answered that question.

*KH: Do you know the answer to that question?*

PC: No. And nobody does. And that's the honest truth. There's something really rather insanitary about quantum mechanics and it's deeply troubling . . .

*KH: It troubles me, I can tell you.*

PC: Well, you are right to be troubled, and physicists are still troubled about this. We know the rules; we just don't know how it jumps, although we do know the way to exactly describe it.

*KH: So how could you guys have come up with something as bonkers as that?*

PC: Well, it is very troubling. Do we believe that there's a reality to nature and it's just that we don't know enough to explain it, or is it that nature doesn't allow us to know the final details, and that physics will just give us the rules to explain the outcomes? At the beginnings of quantum theory, there were two schools of thought around that, and people like Bohr and Heisenberg in the positivist school said 'We can't know about nature', while people like Einstein said 'No, we must know – there is a nature there, there's a God and God does not play dice in this way'.

*KH:* *So Einstein tried to fight it throughout his life, did he?*

**PC:** Yes, he did. Actually Robert Oppenheimer said something rather wonderful about that. He said, 'Einstein put the dagger in the hand of the assassin.' In a sense, Einstein created the ideas of quantum theory but he hated it at the end. It's a fascinating thought – that Einstein was really at the beginning of quantum theory with his photoelectric effect, but later on in his life he really didn't like the way it worked. And he presented a lot of really strange and bizarre consequences of quantum theory to challenge us. He came up with some paradoxes that really showed – in a very dramatic way – how strange quantum mechanics was.

*KH:* *For example?*

**PC:** I can give you a lovely example. Suppose we took a hydrogen molecule that's made of two hydrogen atoms; hence there are two electrons in there. Now, as we discussed, electrons don't like to be in the same state because of this anti-symmetry property.

*KH:* *Hang on, just go back a minute. It has to be hydrogen because . . .*

**PC:** It has to be hydrogen because I want to have two atoms in here. So the two atoms make the hydrogen molecule – they join together. It's $H_2$ in chemistry because there are two hydrogens, and each shares its electron with the other. Now, those two electrons have to be in different quantum states because that's what we learned from the Pauli exclusion principle. And these states are to do with 'up-ness' and 'down-ness' of the electron spin; this anti-symmetry. You can think of the electron being in either an up state or a down state.

Now, we don't know which state the electron is in inside the atom. We've got two electrons in there and we can say that both of them are in a sense partly up and partly down. This is one of the strange things about quantum mechanics.

*KH:* *Hang on, you're losing me.*

209

**PC:** Well, what the uncertainty principle says is that if we go to measure something with precision, we find we'll get that particular result. So if we were to look at whether an electron is up or down in an atom, we'd find that it would be either up or down.

**KH:** *That's a fairly safe bet!*

**PC:** Of course, it's obvious. But before we go to measure it, it's both. Absolutely both.

Now, suppose that hydrogen molecule were to split in two and the two atoms were to head off in opposite directions in the universe. Both those electrons are both up and down. But as soon as we measure one of them and find it's up, it means the other one must be down. It must be. Because we have to have opposite states. So somehow –

**KH:** *No matter how far apart they are?*

**PC:** No matter how far apart they are. So it suggests that the measuring on one side of the universe of the electron being up there has caused the one on the other side of the universe to be down.

**KH:** *You're insistent that we have to have opposite states because why?*

**PC:** Because of the Pauli exclusion principle saying that the electrons have to be in opposite states. But the point of the example I'm giving is that Einstein showed that somehow these two atoms that headed off to opposite directions of the universe had remained connected in a way, so that a measurement on one had affected the outcome of the other. That's like a superluminal connection.

**KH:** *Now, there's no way of proving this?*

**PC:** Well, the experiment's been done!

**KH:** *How?*

**PC:** It was done in Paris in 1982 by a man called Alain Aspect.

*KH: How did he do that?*

**PC:** He fired photons; he didn't do it with electrons, he did it with photons. He fired them in opposite directions from calcium atoms down a very long laboratory – many tens of metres – and he made measurements of the polarisations of the photons at different ends of the laboratory. And indeed he found that the measurement at one end apparently forced the photon at the other end to jump into the opposite state.

*KH: So you're extrapolating that the universe –*

**PC:** Absolutely. The idea works completely: it's saying that everything's connected.

*KH: Yes, well, that's a bit spacey, isn't it?*

**PC:** Well, it is spacey and the technical term for it in physics is 'non-locality', that everything that happens at one place in the universe is affected by everything that's going on anywhere else. This connectedness, of course, gets people speculating in terms of religion and mysticism...

*KH: Yes, I was going to say it rings out and calls God.*

**PC:** I guess so.

*KH: If everything is connected, then who connected it?*

**PC:** That certainly brings us to the idea of what is nature and how nature connects things, and if we think about –

*KH: I mean, sorry to interrupt, but it's an incredibly exciting thought for science. I mean, everyone thinks that science is into compartmentalisation and hard-headed –*

**PC:** It's a myth.

**KH:** *But this is the new-agey-ist thing I've ever heard!*

**PC:** Well, it's pretty new-agey but all of science is about connectedness. I mean, the subject of ecology is how biological ecosystems work. So I don't think the idea of connectedness is unfamiliar in science, but in physics it's very deep: this non-locality property is very obvious in nature and the question of God is interesting, I suppose, because if we believe, as Bohr did, that there are some things that we can't know about atoms or electrons –

**KH:** *Which is what Pascal said about the universe in divine terms...*

**PC:** Essentially ... then we don't need to have an entity that knows about the universe: it's not possible to know. But if we believe, as Einstein did, that really quantum mechanics is somehow imperfect – that there is a knowledge there but we don't yet know it – then you can have the idea that nature had a purpose; that there was some deliberate purpose, some deliberate meaning to physical parameters.

And I guess some physicists often use the term 'God'. It is a curious thing, you know, but it's not good for a young physicist early in their career to refer to God – it doesn't get you published. But if you're a lofty, senior, Nobel-laureate physicist, you can say 'God' and it's sort of OK. But, you know, the physicist's view of God – and we do use this expression in our language; it's the Baruch Spinoza idea of God – goes back to that Spinozan idea in the 17th century: that God is not about a mysterious and omnipresent being; it's about nature itself. Whatever is, is God. We see God manifested in everything – in the stars, water, whatever. That is the natural view.

**KH:** *Well, you've probably already got a formula for it.*

**PC:** Oh ... no. I don't want to go there too much, but there are people, like the physicist–philosopher Paul Davies, who say that 'physics is a more certain route to God than is religion'.

212

*KH:* *It certainly sounds like that.*

**PC:** Well, it's a bold thing to say and I wouldn't want to say that myself, but I think that the idea that there is a nature – that there is a reality that physics allows us to glimpse – means that this nature concept is very powerful. We can interchange it with the word 'God', and sometimes in our seminars and in our discussions we drop the word in. And we know what we mean by it – it is not the same meaning as in ordinary religion.

*KH:* *If you look back in history – I mean, Galileo overthrew the Copernican view ... and all of a sudden everybody's view of everything changed. Well, eventually, once they caught on. Do you think that we're likely to see that happen?*

**PC:** I think we will with quantum mechanics. I don't believe that quantum mechanics is the last word. It works, but it's going to be improved and I think that there's an opportunity for some brilliant mind, some new Einstein, to explain it all to us.

*KH:* *It's not going to be improved before I've understood it properly, is it?*

**PC:** I hope not.

*KH:* *What a shame. I'll have to plod on.*

# BIBLIOGRAPHY

## CHAPTER 1   WHAT IS SCIENCE?

Feynman, Richard. *The Meaning of It All.* Perseus Books, 1998
Gribbin, John. *The Fellowship: The Story of a Revolution.* Allen Lane, 2005
Popper, Karl. *The Logic of Scientific Discovery.* Routledge Classics, 2002
Wolpert, Lewis. 'In Praise of Science'. In *Science Today*, Ralph Levinson and Jeff
    Thomas (editors), Routledge, 1997

## CHAPTER 2   USEFULNESS

Burke, James. *Connections.* Little, Brown & Co, 1978
Burke, James. *The Pinball Effect: How Renaissance Water Gardens Made the
    Carburetor Possible – and Other Journeys.* Back Bay Books, 1997
Callaghan, Paul. 'Breaking the Mold'. Treasury Guest Lecture, October 2006,
    http://www.treasury.govt.nz/academiclinkages/callaghan
Campbell, John. *Rutherford: Scientist Supreme.* Christchurch: AAS Publications, 1999
Djerassi, Carl, and Hoffmann, Roald. *Oxygen.* Wiley-VCH, 2001
Lax, Eric. *The Mold in Dr Florey's Coat: The Story of the Penicillin Miracle.* Owl Books,
    2004
Stolzenberg, Dietrich. *Fritz Haber: Chemist, Nobel Laureate, German, Jew.* Chemical
    Heritage Foundation, 2004

## CHAPTER 3   THE GLORIES OF COLOUR

Greenler, Robert. *Rainbows, Halos and Glories.* Cambridge University Press, 1980
Hewitt, Paul. *Conceptual Physics.* Addison-Wesley Publishing Company, 2005
Lynch, David K., and Livingston, William. *Colour and Light in Nature.* Cambridge
    University Press, 2001

## CHAPTER 4   A LITTLE OF TECHNOLOGY AND THE DIGITAL WORLD

For references on television technology see:
    http://en.wikipedia.org/wiki/Television
    http://en.wikipedia.org/wiki/PAL

http://en.wikipedia.org/wiki/Plasma_display
http://en.wikipedia.org/wiki/Liquid_crystal_display_television
For references on analogue and digital see:
http://en.wikipedia.org/wiki/Analog
http://en.wikipedia.org/wiki/Digital

## CHAPTER 5   SHIMMERING ATOMS
Bodanis, David. $E = mc^2$: A Biography of the World's Most Famous Equation. Berkley Books, 2000.
Campbell, John. Rutherford: Scientist Supreme. Christchurch: AAS Publications, 1999
Cercignani, Carlo. Ludwig Boltzmann: The Man Who Trusted Atoms. Oxford University Press, 1998
Freeman, Ira. Physics Made Simple. Heinemann, 1972
Gribbin, John, and Hook, Adam. The Scientists: A History of Science Told Through the Lives of Its Greatest Inventors. Random House Trade Paperbacks, 2004
Gribbin, Mary, and Gribbin, John. Annus Mirabilis: 1905, Albert Einstein, and the Theory of Relativity. Chamberlain Brothers, 2005
Hewitt, Paul. Conceptual Physics. Addison-Wesley Publishing Company, 2005
Holton, Gerald. Einstein, History and Other Passions. Addison-Wesley Publishing Company, 1995
Lindley, David. Boltzmann's Atom: The Great Debate that Launched a Revolution in Physics. Free Press, 2001

## CHAPTER 6   WHAT IS LIFE?
Davies, Paul. 'Higher laws and the mind-boggling complexity of life'. New Scientist, Issue 2489, 5 March 2005
Hoagland, Mahlon, and Dodson, Bert.The Way Life Works. Ebury Press, 1995
McFadden, Johnjoe. Quantum Evolution: The New Science of Life. HarperCollins, 2005
Schrödinger, Erwin. What is Life? Cambridge University Press, 1967.
Watson, James. The Double Helix. Penguin, 1999

## CHAPTER 7   THE WAY NATURE WORKS
Bak, Per. How Nature Works. Oxford University Press, 1997
Buchanan, Mark. Ubiquity: Why Catastrophes Happen. Three Rivers Press, 2002
Hemelrijk, Charlotte. Self-Organisation and Evolution of Biological and Social Systems. Cambridge University Press, 2005

## CHAPTER 8   EVOLUTION
Dawkins, Richard. The Blind Watchmaker. Penguin, 1991
Dawkins, Richard. Climbing Mount Improbable. Viking, 1996
Dawkins, Richard. The Selfish Gene. Oxford University Press, 1976
Futuyma, Douglas J. Evolution. Sinauer Associates, 2005

# BIBLIOGRAPHY

CHAPTER 9 SEX

Dawkins, Richard. *The Selfish Gene.* Oxford University Press, 1976

Leroi, Armand. *Mutants: On the Form, Varieties and Errors of the Human Body.* HarperCollins, 2003

Ridley, Matt. *The Red Queen: Sex and the Evolution of Human Nature.* Viking, 1993

Sykes, Bryan. *Adam's Curse.* Corgi, 2004

CHAPTER 10 THE GENE, ITS SELFISHNESS AND ALTRUISM

Dawkins, Richard. *The Selfish Gene.* Oxford University Press, 1976

Feynman, Richard. *The Meaning of It All.* Perseus Books, 1998

Sterelny, Kim. *Dawkins vs Gould.* Icon Books, 2001

CHAPTER 11 PSEUDOSCIENCE

Diamond, John. *Snake Oil and Other Preoccupations.* Vintage, 2001.

Feynman, Richard. *The Meaning of It All.* Perseus Books, 1998

Park, Robert. *Voodoo Science: The Road from Foolishness to Fraud.* Oxford University Press, 2000

Rosa, Linda, et al. 'A Close Look at Therapeutic Touch'. *Journal of the American Medical Association.* 279 (April 1998): 1005–1010

Shermer, Michael, and Gould, Steven Jay. *Why People Believe Weird Things: Pseudoscience, Superstition, and Other Confusions of Our Time.* W.H. Freeman & Co, 2002

For more on the story of Emily Rosa, see http://www.abc.net.au/science/k2/moments/gmis9836.htm

For more on homoeopathy and the $1 million prize offered by James Randi, see: http://www.bbc.co.uk/science/horizon/2002/homeopathy.shtml

CHAPTER 12 RADIATION

Collatz Christensen, Helle, et al. 'Cellular Telephone Use and Risk of Acoustic Neuroma'. *American Journal of Epidemiology.* 159 (2004): 277–283

Draper, G, et al. Childhood Cancer in Relation to Distance from High Voltage Power Lines in England and Wales: A Case Control Study. *British Medical Journal.* 330 (2005): 1290. http://www.bmj.com/cgi/content/full/330/7503/1290

*Environmental Health Indicators for New Zealand.* Ministry of Health, 2004. http://www.surv.esr.cri.nz/PDF_surveillance/EHI/EHI_AnnualReport_2003.pdf

http://en.wikipedia.org/wiki/Polonium

Lahkola, Anna et al. 'Mobile phone use and risk of glioma in 5 North European countries'. *International Journal of Cancer, Epidemiology.* 120 (2007): 1769-1775

Lonn, Stefan, et al. 'Mobile Phone Use and the Risk of Acoustic Neuroma'. *Epidemiology.* 15 (2004): 653–659

*National Policy 95.2, Power Lines and Public Health.* American Physical Society, 2005. http://www.aps.org/policy/statements/95_2.cfm

Park, Robert. *Voodoo Science: The Road From Foolishness to Fraud.* Oxford University Press, 2000

Shuler, James Mannie. *Understanding Radiation Science: Basic Nuclear and Health Physics.* Universal Publishers, 2006

Schüz, Joachim et al. 'Cellular Telephone Use and Cancer Risk: Update of a Nationwide Danish Cohort'. *Journal of the National Cancer Institute.* 98 (2006): 1707–1713

## CHAPTER 13    NANOTECHNOLOGY

Feynman, Richard. 'There's Plenty of Room at the Bottom'. Transcript of the talk given on 29 December 1959 at the annual meeting of the American Physical Society, http://www.zyvex.com/nanotech/feynman.html

Joy, Bill. 'Why the future doesn't need us'. *Wired,* Issue 8, April 2000, http://www.wired.com/wired/archive/8.04/joy.html

Kurzweil, Ray. *The Singularity is Near.* Viking, 2005

*Understanding nanotechnology.* From the editors of *Scientific American.* Compiled and with introductions by Sandy Fritz; foreword by Michael L. Roukes. Warner Books, 2002

## CHAPTER 14    THE CLIMATE OF PLANET EARTH

Crichton, Michael. *State of Fear.* HarperCollins, 2004

Intergovernmental Panel on Climate Change http://www.ipcc.ch/

Lovelock, James. *The Revenge of Gaia: Why the Earth is Fighting Back – and How We Can Still Save Humanity.* Allen Lane, 2006

Rees, Martin. *Our Final Century.* Heinemann, 2003

## CHAPTER 15    THE COSMOS

Gribbin, John. *In the Beginning: The Birth of the Living Universe.* Viking, 1993

Rees, Martin. *Before the Beginning: Our Universe and Others.* Simon & Schuster, 1997

Rees, Martin. *Just Six Numbers: The Deep Forces That Shape the Universe.* Phoenix, 2001

Singh, Simon. *Big Bang: The Most Important Scientific Discovery of All Time and Why You Need to Know About It.* Fourth Estate, 2004

## CHAPTER 16    QUANTUM MECHANICS

Al-Khalili, Jim. *Quantum: A Guide for the Perplexed.* Weidenfeld and Nicholson, 2003

Bryson, Bill. *A Short History of Nearly Everything.* Doubleday, 2003

Davies, Paul, and Brown, J.R. *The Ghost in the Atom: A Discussion of the Mysteries of Quantum Physics.* Cambridge University Press, 1986

Hey, Tony, and Walters, Patrick. *The New Quantum Universe.* Cambridge University Press, 2003

Penrose, Roger. *The Emperor's New Mind.* Oxford University Press, 1989

Singh, Simon. *Big Bang: The Most Important Scientific Discovery of All Time and Why You Need to Know About It.* Fourth Estate, 2004

# INDEX

219

# INDEX